Cosmic Sense, Common Sense, and Nonsense

The Philosophy of Science Applied to Modern Physics

Table of Contents

All images for cover courtesy of Wikimedia.

By Douglas E. Reinhardt
Ph.D., UNC-Chapel Hill

Key Subjects
Astronomy, Big Bang, An Alternate Physics, A Skeptical Approach to Modern Physics, Philosophy of Physics, Epistemology, Anti-relativity, Quantum Mechanics, Quantum Mysticism, Scientific Realism, Metaphysics, Cosmology, Critical Thinking, Metacognition, Research Methods, Logic, Rational-Empiricism, Logical Positivism

Key Question:
Does modern physics use a different kind of logic or thought process from that which is used in other sciences?

Copyright©2018: First Edition
by Douglas Reinhardt
All Rights Reserved
IozilB

For those who wish to skip the introductory information and get right on with the reading, click this link to the **Table of Contents.**

COSMIC SENSE, COMMON SENSE AND NONSENSE

About the Author

I have enjoyed a life-time interest in the physical sciences. When I entered high school, my favorite subject was physical science, and I had planned to make a career in science or engineering. Nearing graduation, I won a scholarship to study engineering, but by that time, I had changed my mind, and I wanted to become a teacher. While in college in the 60s, I became intrigued with the social revolution that was taking place and decided to major in the social sciences. I pursued a major in anthropology with a minor in sociology. I chose anthropology because it allowed me to study social science (Social and Cultural Anthropology) and biological science (Physical Anthropology). However, I never lost my interest in physics, and astronomy became my hobby - complete with telescope and camera attachment. When Carl Sagan came out with his *Cosmos* series in the late 70s, early 80s, my interest in physics was rekindled. I watched all his videos, read the companion book and watched his movie, *Contact*. I was enthralled with the seemingly otherworldly theories of relativity and accepted such concepts as time dilation on faith – faith in the authority of my hero, Carl Sagan and other scientists.

Fortunately, I had several friends who kept me in touch with developments in physics and astronomy, and then I had an encounter with a student who had studied physics from his childhood, and it seemed, he had read almost every popular book on physics ever written. In my many conversations with him, I began to want to know the evidence behind the seemingly exotic theories of relativity and quantum mechanics. As I delved into this evidence, I became more and more skeptical of the way language and logic were used to describe non-classical phenomena.

My training in anthropological linguistics had taught me that the way we perceive the world is largely influenced by the language we speak, and the logic embedded in it. I was particularly skeptical of the way relativity treated abstractions such as space and time as physical entities. I was told that the theories of relativity and quantum mechanics were counterintuitive and could not be understood with common sense rationality. In my efforts to understand these exotic phenomena, I was fortunate to be tutored by a retired physics professor for a semester. Although he tried to liberate me from my classical tendencies, I became more skeptical as I learned more about the evidence and line of reasoning that led physicists to the bizarre conclusions of modern physics.

After retirement from teaching, I decided to follow my passion and study physics on my own. After reading several books on the subject, participating in blogs and message boards, discussing my ideas with friends who believe in the current theories of physics, I decided to write my own book, challenging the language, logic and interpretation of experimental results that support these outlandish theories. Fortunately, I found several authors who had similar objections to the thought processes in modern physics. So here I am, with my first book on the philosophy of physics.

I started this book with a neat outline as to how I was going to organize the subject in air-tight categories. As I began writing, however, I discovered more and more interconnections among topics that made it difficult to stick with my original outline. In proofing my work, I found that I write much like I taught. I state my points in several different ways with quite a bit of repetition and redundancy in hopes that if the student (in this case the reader) didn't catch my point the first

COSMIC SENSE, COMMON SENSE AND NONSENSE

time, s/he could not miss it on the second or third go-around. If you find some parts too repetitive, just say "OK, I got it" and speed read on to the next point. I hope you will find my book challenging to your established beliefs about the "nature of nature."

My only fear in presenting my dissident views of Modern Physics is that some physicists may find a technical error in my interpretation of a theory and discount the whole thing. I would ask the reader to consider the work as a whole and decide in the main as to whether Modern Physics violates the philosophy of science and the scientific method in some of its speculative theorizing. Is it true that much of physics today is based on metaphysical mathematics and even mysticism (rather than empiricism) as some physicists, such as Capra and Bohm, have already acknowledged?

COSMIC SENSE, COMMON SENSE AND NONSENSE

Dedication and Acknowledgements

I dedicate this controversial book by an author who has ventured far outside his field to my loving wife, Peggy, who has been so tolerant and understanding of my need to make a statement on my perception of the "state of science" particularly in the field of physics. In our post-retirement years when folks are supposed to be sitting in rocking chairs, traveling and socializing with friends and family, she has supported me in every way despite my taking time away from her to pursue this passion.

I am also indebted to my science-oriented friends who have continued to stimulate my interest in hard science over the years. Those who argued for the standard theories of physics and resisted my classical tendencies have made me sharpen and fine tune my arguments for an alternate way of viewing the "nature of nature" and bring the "physical" back into physics.

And, to all my family, who may have felt neglected because of my solitary pursuits in reading, thinking and writing, I appreciate your support and understanding.

COSMIC SENSE, COMMON SENSE AND NONSENSE

Table of Contents - toc (page numbers apply only to print book)

COSMOLOGY: RELATIVITY VS. QUANTUM MECHANICS - p. 6
Big Bang Cosmology and Expansionist (Inflation) Theory - p. 6
The Big Bang: A Creation Story (Myth) about How the Universe Came from Nothing – p. 10
Hawking's Contribution to the "Nothing Theory" (Hawking Radiation) – p. 20
Dualities of Mind and Dualities in Nature – p. 25
Susskind's Refutation of Hawking Radiation and Information loss in a Black Hole – p. 26
Gravity: A Push or Pull, Attractive or Repulsive? – p. 31
The Shape of the Universe: Flat or Soccer Ball Shaped? – p. 34
Seeing the Big Bang - p. 37
Alternatives to the Big Bang: Steady State and Quasi-Steady State – p. 42
William C. Mitchell's Recycling Universe Cosmology (RUC) - p. 46
Physics, the New Religion??? A Multiverse Tied Together with Strings – p. 49

SUMMARY – p. 50

References-p. 51

COSMIC SENSE, COMMON SENSE AND NONSENSE — Big Bang & Inflation

COSMOLOGY: RELATIVITY vs. QUANTUM MECHANICS

Cosmology is wedding the very small to the very large.

If the law of conservation of mass and energy is correct, then the universe has to be eternal. If matter and energy cannot be created nor destroyed, it must have always existed and will always exist.

Modern cosmology (and astronomy) is largely seen through the lens of relativity, particularly General Relativity which is a theory of gravity, or perhaps "non-gravity" gravity. Lawrence Krauss, trained in quantum mechanics and cosmology, attempts to wed relativity to quantum mechanics to show how the universe came into being. Krauss has written a book entitled *A Universe from Nothing* (2012) in which he attempts to show how the universe could have risen from a "deeper nothing" using the theories of relativity and quantum mechanics. If one examines the derivatives of relativity, fantastic extensions of the theory arise when applied to outer space. An example of a fantastic derivative is the wormhole where space and time travel are made possible by folding up space like a piece of paper and boring a hole through it. Other fantasies involve contracting space by traveling near light speed, expansion of space under the repulsive force of dark energy (which is driving the galaxies apart faster than light yet not violating the speed limit of light), space and time being created in the Big Bang, a geometric singularity of spacetime that existed before the Big Bang, *ad infinitum*…

Big Bang Cosmology and Expansionist (Inflation) Theory

The expansionist or inflation theory that support Big Band cosmology leave much to be desired in terms of language and logic. The Big Bang and its derivative, expansionist theory, draw heavily from Einstein's General Relativity. The basic fallacy of all these theories is semantic and linguistic in nature. The key word which is the underpinning of all these fantastic theories is **space**, which is treated in modern physics as a thing with physical properties and physical structure. If space is a form of matter, then it is a redundant concept – just call it intervening matter and be done with it.

Actually space is the opposite of matter – it is a volume where there is ideally no matter and energy. It is thus inversely related to matter and energy. In a given volume, the more matter there is, the less space there is. Conversely, the more space there is, the less matter there is. In other words, we are dealing with density which is defined as mass/volume or matter/space. The higher the density of a volume, the more matter there is, and the lower the density of a volume, the more space there is. Therefore, space exists where there is not maximally dense matter as in a Black Hole. To say that space (vacuum) is a thing is tantamount to saying that nothing is something – an egregious error in elemental, binary logic. Something cannot be 0 and 1 at the same time. Finally, I found someone who agrees with me on the nature of space. William C. Mitchell (2002) says "But the belief that space is expanding is folly. Empty space is nothing, and nothing can't expand (p.234)."

Another fallacy relating to language is that **space and time were created in the Big Bang**: I have heard a number of cosmologists make the claim that space and time didn't exist before the

Big Bang. However, when they say that there was a singularity or point where all the matter of the universe was compressed prior to the bang, then they are admitting that time and space already existed. However, since I see time as human creation designed to standardize various rates of change, time, then, is an abstraction of our creative brains. Change and motion may have been created by the Big Bang, but it took humans to find a standard by which to measure all these various rates of change and call it time. However, if space is nothing, "nothing" does not have to be created at all. Space (nothingness) would exist before the Big Bang – it is a given. This is an obvious problem of language and logic.

It may sound prosaic and pedestrian to quibble over the definition of space, but if space is not a physical structure, then the theories of Relativity and all their derivative theories from time travel to the Big Bang and Expansionism reduce to absurdity. Obviously if space is not a thing, then it cannot be expanding. Matter may be spreading out in space producing a less dense universe, but this language implies something quite different. Let's review some of the language and logic by the chief proponent of Expansionism, Dr. Alan Guth.

Space is expanding in all directions between galaxies faster than the speed of light. This idea derived from GR is that galaxies are not flying apart from each other in the emptiness of space, but that space (nothingness) is increasing between them – some say faster than light. Kaku treats us with another Zen-type koan to explain this in "Something from Nothing": "Dark energy is vacuum expanding – the energy of nothing – even vacuum expands to hyperspace – the energy of nothing is pushing the universe apart." When questioned about the violation of the speed limit of light in saying that space is growing faster than light, Kaku in another video says: "Space is nothing, and *nothing* can go faster than light." Obviously if space is growing between galaxies faster than light, then space is pushing galaxies apart faster than light, and Einstein says it would take infinite energy to accelerate mass to the speed of light, so mass carried by expanding space would indeed be violating the speed of light.

For Kaku and others who say that it is space that is expanding faster than light rather than galaxies moving through space away from each other at light speed, consider that if an earthquake creates a rift in the earth's surface, the objects on either side of the rift are moving away from each other as space is created between them. The other problem with this faster-than-light expansion idea is that if distant galaxies are sailing away from us faster than light, then we would never see them because we would be out-running the light, thus proving that the speed of light is not the same for all observers and is therefore relative rather than invariant.

Still another problem arises from this faster-than-light expansion. Recall that when objects approach light speed, space has to contract as time slows down. Now we have a contradiction because as space is expanding and driving matter apart faster than light, relativity is causing space to contract and therefore to reverse the process. So, which is it, is space expanding because of dark energy or is space contracting because of high speeds? If you say both, then what is the net speed of contraction or expansion?

Furthermore, if space is expanding faster than light and pushing the matter outward at the same rate, the matter would become infinitely massive, and it would require infinite energy to push it because Einstein's speed limit would have been violated. The patch for this argument is that it is not the galaxies that are flying apart faster than light, it is the space between them that is expanding faster than light and the galaxies are at rest in this space. Of course, if the galaxies are being carried faster than light, they, too, are indeed going faster than light relative to space. If my car is carrying me at 70 mph, then I am traveling 70 mph relative to the road (earth's surface).

Before the Big Bang, there was a singularity of spacetime: Some cosmologists speak of highly compressed matter similar to a black hole, others speak of a singularity. The idea of a singularity in spacetime is a geometric concept, not a physical entity. A point is defined in geometry as having zero dimension – therefore it is an ideal, Platonic concept, not a physical reality. If you say "something came from nothing", then you are giving credence to the Biblical concept of *ex nihilo*. However, that seems to violate the law of conservation of mass and energy. Here's a riddle or koan from Michio Kaku's YouTube video, "**Something from Nothing**."

* Total matter in universe is positive - total energy is negative (because gravity is negative) = 0
* Total positive charge - total negative charge = 0
* Positive spin of galaxies - negative spins of galaxies = 0
* Therefore the universe is zero because total matter, total charge and total spin equals zero, so the universe could have been created from nothing.

Although I admire Michio Kaku for his energy and charisma in popularizing physics, I don't think this kind of logic is worthy of him. Just because the charges of the universe balance out to zero does not mean that the matter (some of which is positive and some of which is negative) does not exist, that is, it is nothing. If there is an electron with a negative charge of 1 and a proton with a positive charge of 1, there is zero net charge, but you don't have zero matter or mass – you still have a proton and an electron. The same is true of matter and energy and spin. Having an equal amount of matter and energy does not mean that matter and energy do not exist, and if half the galaxies spin in one direction and half in the other, it doesn't mean that galaxies don't exist or came from nothing. How this all adds up to the idea that the universe could be made from nothing and how the universe might be "nothing" is beyond my logic and language to describe. I would characterize this as mystical thinking.

However others see the singularity not as a dimensionless point but as all the matter and energy of the universe compressed into a very small space. Hawking (2014) says that the initial singularity was the gravitational singularity of infinite density thought to have contained all of the mass and space-time of the Universe. If the singularity did contain all the matter in the universe compressed into a small space (but not a single point with no dimensions), then the question becomes: "How did it get in that shape to begin with." Highly compressed mass would suggest that there was a big crunch before the bang - a theory that has been largely abandoned in light of apparent accelerated expansion and the lack of visible matter and dark matter to stop the expansion form going on forever. However, the universe might be like a supernova which, when it becomes tightly compressed, explodes in a blaze of glory.

There are at least three highly dense objects in space that have been theoretically identified: the singularity which existed before the Big Bang, black holes, supernovae and neutron stars. The question becomes why some high-density objects explode and others do not. Why did the singularity of Big Bang fame explode, and why do supernovae explode leaving a core that is highly compressed neutron star? On the other hand, black holes do not explode nor do neutron stars. One would conjecture that if the inward pull of gravity is greater than the outward pressure of radiation in the form of heat, then the object will not explode even though it might leak radiation as black holes supposedly do. If on the other hand the outward pressure of radiation is greater, the object will explode breaking the bonds of gravity and the electro-magnetic bonds and strong force that holds matter together thus releasing radiation in the form of subatomic particles. Of course, a singularity is too simple a concept if one means a point with zero dimensions. Matter, no matter how compressed, exists in three dimensions. Scientific American astronomy editor George Musser agrees that "singularities are an idealization" (2003).

The Empty Shell Universe. Big Bang theory indicates that space is expanding in every direction from the singularity before the Bang, but there is no center to the universe. The universe is like a big bubble and each point on the surface of the bubble appears to be the center from its own perspective. To begin with, a bubble would have a geometric center since it is spherical in shape. Perhaps GR-type cosmologists mean there is no mass left in the center and what we have is a hollow ball. If this is so, then all galaxies would be floating on the membrane of the bubble, and we, in the Milky Way, would see galaxies relatively near on the surface of the bubble and others would be extremely far away on the opposite surface of the bubble. And, looking in the opposite direction from the surface of the bubble, we would see nothing but the blackness of space. Obviously, this is not what we see, and this concept derived from this GR-derived notion of space expansion is fallacious. More on the shape of the universe below.

The Big Bang: A Creation Story (Myth?) about How the Universe Came from Nothing by Lawrence Krauss and the Big Bangers

Allow me to tell the creation story according to the high priest of nothingness, Lawrence Krauss (2012), and his colleagues on the miracle of getting something from nothing. First, we had the Theory of Everything (TOE) and now we have the Theory of Nothing (TON), but, according to Krauss, the Theory of Nothing is the Theory of Everything because it was *nothing* that made *everything*. Here's how the story goes.

1) In the beginning, there was nothing – except for a singularity in spacetime* (a geometric point with zero dimensions – in a word, "nothing").

2) Suddenly, the singularity of nothing got really hot and began to expand very quickly creating even more nothing (space). This expansion of nothing (space) was caused by dark energy – "the energy of nothing". You see, even though there was already nothing before the Big Bang, a "deeper nothing" had to be created so that something could come into existence. This deeper nothing is even more nothing than zero nothingness – it is negative nothing. According to some nothingnists, deeper nothing, which is like negative mass, can cause negative curvature of the nothingness of space, but Krauss says the data suggest that the void of space is flat so that the "somethings" will travel straight out from a non-center and burn out eventually returning to nothing.

3) However, from the nothingness of space, virtual particles began to become "something" temporarily, but they had to become nothing again very quickly so as not to violate the law against getting something out of nothing (conservation law). Since these particles paid back the energy they borrowed from nothing, their check balance was zero, so the particles really amounted to "nothing" after all and didn't violate any laws.

4) The way all this happened was that the virtual particles were created as twins. One twin was matter and the other was anti-matter and they were mirror images of each other. Since one had a negative charge and the other had a positive charge, they found each other attractive, but when they met, they annihilated each other and went back to being nothing (except radiation). However, for some ungodly reason, there was more matter created than anti-matter so from the left-over matter, the universe became materially real.

5) So somehow the virtual particles of nothing became real particles of "something", embezzling energy from the nothing of space and not paying it back right away. On judgment day, however, when the universe burns out, these real particles which became hard matter will have to return to nothing – their karma for cheating the law of conservation.

5) But in the meantime, the real particles coagulated and formed large blocks of matter which found each other very attractive (gravity) and so they formed islands of galaxies containing stars, planets, moons, nebula and the like. However, since there was not enough visible hard matter to hold the galaxies together, they were helped out by invisible "dark matter" to resist the dark side of the force (dark energy) which would tear them apart by expansion if not pulled together by dark matter and regular matter. So dark matter and dark energy are in a tug of war pulling matter in opposite directions, but eventually the nothing (dark energy) will win over something.

6) However, the large seas of empty space surrounding the island galaxies continue to expand creating more and more nothingness under the influence of the "energy of nothing" (the dark side of the force or dark energy); thus driving the island galaxies farther and farther apart faster than light. The expansion is so fast that distant galaxies appear red like a setting sun. Now the founder

of nothing theory – one named Einstein – said that if any "something" goes as fast as light, it will acquire infinite mass (lots of something) and require infinite energy; thus "something" would get piled higher and deeper (PHD) and replace the nothingness - and that would kill the nothing theory. Nothin' doin', says Krauss, it's not the galaxies that are moving faster than light; it's space (which is carrying the galaxies) that is expanding faster than light, and since the galaxies are at rest in space and just being carried along by spatial expansion, the galaxies are not violating the speed limit of light. See, there's nothing to it – the galaxies are just along for the ride – they aren't really moving. *Nothing is moving because space is nothing and nothing can go faster than light.* However, in the real physical world that we know, if we put a spacer between two objects, the spacer causes the objects to move apart in space, but ordinary logic doesn't apply in the realm of the surreal and super-real.

7) But wait there's another problem. If the nothing of space is expanding faster than light, then light would never reach us from other galaxies, because space would be out-racing light. It would be like the poor swimmer, swimming upstream against a current that is flowing faster than she is swimming – she keeps going backwards and never gets to her destination. But clearly we do see distant galaxies, so, "Houston, we have a problem," and we need to have a séance with Ptolemy to get another epicycle-type hedge to rescue us from this paradox to prevent the paradox from becoming a contradiction.

8) Despite gravity and dark matter's struggle to hold matter together, the energy of nothing (dark energy) will eventually win and rip the galaxies apart and the stars will burn out, reducing matter to radiation and returning all the "something" to the cold, dark void of nothing to settle the score and balance all accounts to zero, so that "nothing" reigns supreme once more in a non-universe.

9) This *scientific* creation story replaces the old, superstitious story in which an intelligent designer created something out of nothing. The old story is very unscientific, but the new story is hard, rational science based on natural law and mathematics.

10) However, one of the high priests of the new story named Kaku says it is "undecidable" as to which story is true. Being undecided is in opposition to the high priest of secret knowledge named Krauss who says the new story is backed up by strong scientific evidence, and the old story is a fairy tale and definitely false.

11) However, Krauss cannot explain how the universe made something out of nothing and why you can't make nothing out of something without violating the law of conservation of mass-energy. This law of conservation would dictate that the universe is eternal, because, since you can't get something out of nothing, everything always was and always will be. Krauss continues to argue that you can get something out of nothing as long as you pay it back in a timely manner. Since the amount of something borrowed from nothing is equal to the amount of something repaid to nothing, the universe has a balance of zero which is, of course, "nothing".

12) To explain how the universe got on welfare and received a free lunch from nothing, Krauss called on Kaku, a string theorist whose theory is hanging on by a thread, and Guth, a fellow Banger who has inflated the theory of the Big Bang. They say that the

* *Total matter in universe is positive minus total energy which is negative = 0*
* *Total positive charge minus total negative charge = 0*
* *Positive spin of galaxies minus negative spins of galaxies = 0*

Therefore the universe is zero because total matter, total charge and total spin equals zero, so the universe could have been created from nothing. Got it? But wait, there is a lot more matter than anti-matter creating a non-zero sum, so matter can't be reduced to zero and just go out of existence by being annihilated by anti-matter. What shall we do with this inequality in nature and the stubborn persistence of matter? Then there's the problem that dark energy makes up 70% of the universe, dark matter makes up 26% and regular matter makes up only 4%. Well, we have yet another inequality that doesn't balance out to zero, but I'm sure Krauss and the bangers have another Zen koan to get us back into the zero-sums game.

13) The free lunch, however, is mostly jello which Krauss and company nail to the wall with great dexterity. For example, when one proton which is positive combines with one electron which is negative to create an hydrogen atom, there is a net charge of zero; therefore the hydrogen atom doesn't exist – because it is zero.

14) One of the rules of the higher nothing theory is that common sense and rationality are not allowed, so we must accept the equivalence of "zero net charge" and "zero existence" without question. To paraphrase Feynman – nobody understands this stuff, we must accept it on faith in the mathematics. It really requires a quantum leap of faith to get it. If you don't get it, then you don't know nothin' 'bout nothin'.

15) When you have divested yourself of all rationality, you should be convinced that "nothing" is not really "nothing" but is actually quite "something". In the nothingness of space, there is information about natural law, but it does not have any material reality and doesn't require energy to carry the information - unlike modern communication which requires radio waves to carry our voices and images across space. When you achieve this sudden flash of enlightenment, you have achieved nervana (similar to nirvana). Nervana can be achieved only after emptying your mind of all "something" – and nervana comes after lengthy and diligent meditation of contemplating your navel or zero. Only then can you hear the sound of one hand clapping in the emptiness of the universe.

15) So in the end, the universe will return to nothing when all debts are paid, and all balances are zero. But maybe another bang could happen if something could bang out of nothing like before and a new universe be born.

16) This is Krauss's story - and he's sticking to it - on how the universe came from nothing and will eventually return to nothing. In the end, his theory is "much ado about nothing."

17) Shakespeare may have more to say about "nothing:" "(The Big Bang) is a tale told by an idiot, full of sound and fury – signifying nothing." Now I am not calling Krauss an idiot because he is obviously a very intelligent man, but I think he is stuck inside the box of modern physics, and it is difficult for him to think outside that box with all its questionable assumptions and mystical logic. However, I do believe the "sound and fury" of the bang may indeed signify "nothing." The bang tells us the universe created itself.

* Inflation theorists such as Alan Guth have changed their minds about a singularity being a dimensionless point in spacetime. Now, they appear to be saying that there was highly-compressed matter before the Big Bang so that the universe may not have come from nothing after all. However, this admission of the existence of something before the Big Bang means that space, time, matter and energy were not created by the bang implying that the universe is eternal rather than having a sudden beginning in time as originally stated. Now the concept of "eternal inflation" has caught on which is said to create multiverses, so the idea of eternity is coming back into vogue – if not an eternal universe, then an eternal multiverse. Another theory of the cause of the bang comes from quantum physics which says there are lots of energy in empty space, and every once in a while, there is an accidental concentration of this energy called a quantum

fluctuation. A large quantum fluctuation could have caused the bang, and the bang spread throughout space as the vacuum energy began to heat up and expand and create particles from energy.

Krauss believes in the non-material while rejecting the spiritual
Actually Krauss, despite all his anti-theistic rhetoric, asserts a strong belief in non-material reality even though he does not call it spiritual. Let me count the ways:
1). "Nothing" is non-material; thus getting something from nothing is obtaining something from a non-material reality. There are many non-material "nothings" in his argument. There is the nothing before the bang, the nothingness of space, the energy of nothing (dark energy).
2) To get something from nothing, the "nothing" had to have information for constructing "something" even though there was no mass or energy to carry the information. However, to the rational mind, it is difficult to conceive of disembodied information without a carrier such as electro-magnetism.
3) As a quantum physicist (before becoming a cosmologist), Krauss must believe in consciousness (a non-material reality) which affects physical reality. Recall that quantum physics indicates that observation (consciousness) by the experimenter determines the outcome of a quantum event, i.e., whether the cat is dead or alive or whether a quantum is manifesting as a wave or particle. Consciousness is associated with matter and energy, but it is in itself non-material.

Getting around the God problem, the first cause and prime mover

Here again, Krauss performs verbal acrobatics, including back flips, to get around the problem of a first cause. Since he believes in the Big Bang which means that the universe has a definite beginning in time, he is forced to deal with the prime-mover problem. Ironically, Krauss shares the concept of the universe (having a beginning and end) with Christians whom he often rails against. However, since both worldviews posit a beginning, it seems logical that the universe had to have a prime mover. Buddhists get around this problem by saying that the universe is eternal. Buddhists reason that everything has a cause, so since there can be no object or person without a prior cause, there must be an infinite regression of causes into the infinite past with no beginning and an infinite progression of cause and effect so there is no end. But since Krauss and Christians believe in a definite beginning, then there had to be a first cause to get it all started. Christians believe that the first cause was personal, i.e., God, who, in turn, is eternal having no beginning nor end. The buck stops here. Krauss on the other hand believes in an impersonal universe so the prime mover or first cause was information about natural law which, like God, has no material existence. Krauss, in true Zen fashion, says he gets around the first-cause problem because the universe came from nothing and therefore "nothing" cannot be said to be a cause or have a cause because it doesn't exist. But, of course, the one hand clapping gets even louder when he says that nothing is not really totally nothing because there is information on natural law in empty space. But that assertion would lead to the conclusion that natural law, like God, always was and had no beginning propelling him back to the Buddhist view of an eternal universe.

But wait, Krauss has yet another Zen koan for us. Since there is a multiverse (again a speculation) and some universes within the multiverse have different variables and constants (natural laws) from ours, then natural law is universe-specific and is therefore not eternal since each universe has a beginning and an end and thus an end to the laws that put it together. Now Krauss is caught in a trap from which he cannot escape except through mysticism. If there is an irresistible force that somehow drives empty space to produce a multitude of universes, then there is a higher natural law that dictates what constants and natural laws a given universe will have. In our universe, the Planck constant may be a little different from that constant in another universe, and two quarks may make a quasi-proton instead of three quarks that make our proton, etc. It is rather like parents, who, from the same gene pool, produce children having different combinations of DNA with perhaps some mutations thrown in, for example, this kid got brown eyes, the other got blue eyes, and the other was an albino and got pink eyes etc. Krauss weakly counters this argument of a higher natural law that produces a variety of universes by saying that the universe arose as an accident, implying that no eternal natural law was necessary, so the universe must have popped into existence without a prior cause. In other places, however, Krauss has said that empty space contains natural law information, so he tries to have it both ways – depending on which way suits his argument at the time. Michio Kaku has a much more rational point of view on the subject in the view of this author. He says the whole issue is "undecidable" as to whether the universe was created by an intelligent designer or arose via impersonal natural law. I say that this genesis issue is a matter of personal faith and lies beyond the realm of science. It takes as much faith to believe one view as it does the other. There is just no way of skirting the prime mover problem in a universe that had a beginning despite Krauss's Herculean efforts.

As a post script to the prime-mover issue, the term "multiverse" is a misnomer. Since the "universe" means everything that exists, then there can be only one universe ("uni" means one). If there are other super-galactic systems like the one we live in, then they should be called "subverses" which combine to make up the universe. This may seem like semantic quibbling, but the thesis of this book is that imprecise language (such as the definition of space and time in physics) has led to some fantastic and untenable theories.

Furthermore, Krauss's argues that the universe is zero because total matter, total charge and total spin equals zero which means that the universe could have been created from nothing. However, he ignores the fact that there is a lot more matter than anti-matter creating a non-zero sum, so matter can't be reduced to zero and just go out of existence by being annihilated by anti-matter. How does Krauss get around this non-zero sum game that the universe plays with us? Can he hatch enough anti-matter from the quantum foam to annihilate all the matter in the universe? The problem is that anti-matter and matter do not obliterate each other, they transform their material substance into energy, and if Einstein is right, matter and energy are interchangeable – so nothing is destroyed in the sense of going out of existence. So, Krauss's verbal smoke screen is blown away.

Krauss's "Nothing" theory is no better than God Argument"
1) If you say that God is the first cause of bringing the universe into being, then you must explain the origin of God, that is, what caused God?
2) If "nothing" is the first cause, there is no need to explain the origin of nothing because nothing doesn't exist (according to Kraussian logic). But, of course, one has to explain the force and natural law that brought something out of nothing, then one has to account for the origin of that force, so we are back to the infinite regress dilemma that Krauss ascribes to the God problem. In either case (God or nature) there is either an infinite chain of causation or

> there is something which existed forever that was the first cause that set off a chain of existence.

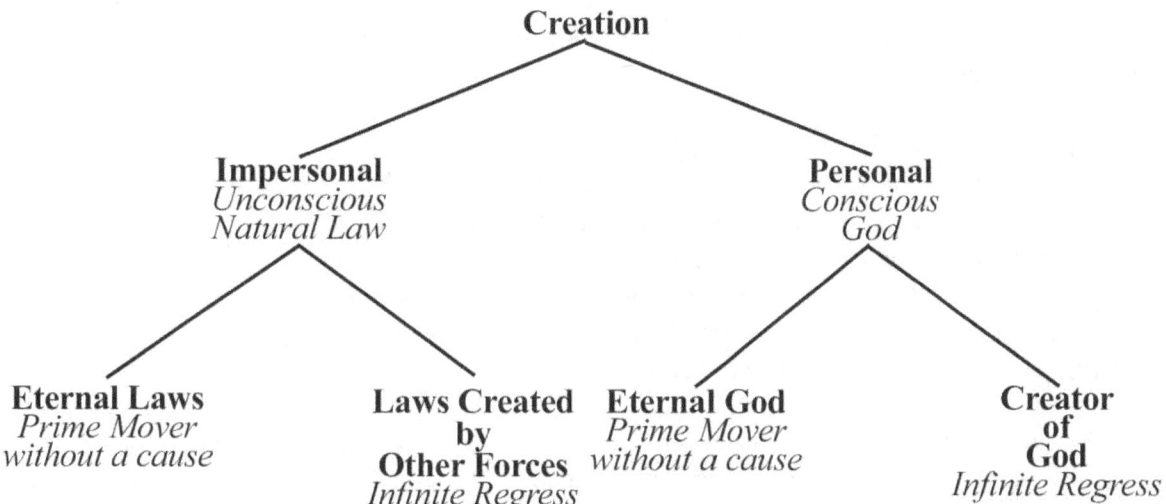

This graphic demonstrates that whether you believe the universe is personal or impersonal, you hit the same logical impasse. Either there was a prime mover and the universe has a definite beginning in time, or there is an infinite regress of causes and the universe is eternal. Thus the idea of no-God presents the same logical problems as the idea of God.

The Laws of Thermodynamics and Cosmology

The three laws of thermodynamics have many implications for cosmological theory.
1>The first law of conservation of energy in a closed system has implications for the Big Bang theory. If, as Einstein said, energy is equivalent to mass, then we could restate the law that neither mass nor energy can be created or destroyed – although they can be interchanged as indicated in the formula: $E=MC^2$. Assuming that the universe is a closed system, then we would have to say that the Big Bang could not have arisen from a singularity in spacetime with zero dimensions. That would be getting mass and energy from nothing. We would have to infer that there was an equal amount of mass and energy before the Big Bang as there was after the bang. Also, we could not say that particles spring up from the nothingness of space. Physicists try to get around this by saying that pairs of particles (one matter and the other antimatter) spring out of nothing, but they don't violate the conservation law because they attract and annihilate each other and return to zero. However, inverse-matter particles do not annihilate each other in the sense of returning to nothing - they convert mass to energy and remain as radiation (photons). Thus, the conservation law is violated if the claim in that particles arise out of nothing. However, to claim that particles condense out of the background energy in space would not violate this conservation law, but that would mean that this energy was present at the time to the Big Bang. Furthermore, this argument doesn't explain how more matter was created than antimatter.
2>The second law of entropy (heat and particles, mass and energy dissipate to a state of equilibrium and become disorganized) supports the idea that the universe will experience heat death and become totally disorganized as a sea of expanding radiation. This dissipation and

disorganization of the universal system is exacerbated by the fact that the universe is expanding rapidly thus scattering mass-energy at a faster rate than if the universe were in a steady state. However, Boltzmann amended this law by saying that occasionally, because of random fluctuation of particles, there can be a reversal of entropy and a movement toward recovering order. If the Big Bang was such a fluctuation, then it would seem possible that the entropy of the universe could be reversed.

Regarding inflation, it is said that the expansion of space (caused by dark matter) is between galactic clusters, not within galaxies and galactic clusters where dark matter holds these large systems in place. Assuming that the gravity of matter and dark matter are as strong as dark energy, then entropy seems reversible in these rather closed systems of galaxies. For example, a star such as our sun, as it undergoes entropy, produces order and complexity on a planet such as earth so that entropy in one subsystem produces extropy* in another subsystem at least temporarily. We also see that as massive stars undergo entropy as they burn out and become super novae, they fuse lighter elements to heavy elements which can be used to support life in another solar system such as ours. Some physicists may contend that dark energy, making up 70% of the universe, will eventually overcome the grip of gravity and rip the galaxies apart and lead to ultimate entropy, and, if so, entropy will be the winner. Further, if galaxies are leaking some mass-energy despite the grip of gravity, this would make ultimate entropy possible. However, entropy does not explain how the universe was so extropic* in the beginning with the Big Bang. Some explain this extropy* as random fluctuation and others by intelligent design, but, as Kaku asserts, this question is undecidable.
* Extropy as used here is defined as the opposite of entropy and is not meant as support for the biological theory of extropy. Perhaps negentropy would have the same meaning as extropy.
3> The third law of thermodynamics states that the entropy of a system approaches a constant value as the temperature approaches absolute zero. This would mean that the matter and energy are evenly spread (in equilibrium) as the universe approaches the ultimate freezing point where that is no more particle motion. Achieving absolute zero would seem to violate the first law that energy cannot be destroyed. If this is so, the universe would become a cold place but not frozen. However, a random fluctuation, according to Boltzmann could reverse this process because everything that can happen will happen given enough time.

Random Fluctuation vs. Intelligent Design: Where does complex order come from?

The question of creation of life and consciousness is as old as human consciousness itself. There are three major answers to this question: 1) Order was created as a random fluctuation. In other words, order comes out of chaos by chance. 2) Order is inherent in nature itself in an impersonal, unconscious universe. 3) Order was created by an intelligent designer who is conscious and personal.

Random Theories: Let's look at the random theories first – the idea that low entropy life came from a universe regressing toward high entropy by chance. 1) Krauss has informed us on the theory of how the *Big Bang* could have arisen as a random fluctuation in space. This is based upon the idea that virtual particles can randomly bubble up out of empty space. Predicated upon this idea of a quantum foam, the universe itself could have accidentally arisen from "a deeper nothing" as Krauss puts it. 2) According to the *Boltzmann's Brain* thought experiment, life and consciousness could have arisen from a random fluctuation of particles in a universe that is running down toward entropic death (disorganization and thermal equilibrium). Boltzmann's idea is actually a contradiction to the second law of thermodynamics which indicates that in a closed system, entropy flows down a one-way street from high organization to low organization. Considering the universe to be a closed system, this idea begs the question of how the universe got into a low entropy (high organization) state to begin with. Of course, Krauss has told us that the Big Bang was a random fluctuation that created low entropy. However, since the Big Bang created a low entropy universe, things have been running down, so how is it that complex life with consciousness has arisen in this system where everything else is running in the opposite direction, i.e., toward greater entropy and disorganization? Boltzmann says that the second law doesn't always follow that one-way street, and sometimes it can reverse itself and create order in a random event. Given enough time, anything that can happen will happen, therefore random fluctuation can create a disembodied brain that is self-conscious. This lone, unembodied brain is much more likely than a brain attached to a superfluous body; thus, there are more solo, unembodied brains in the universe than embodied brains found in aggregates called societies. So, if Boltzmann is right, the second law of thermodynamics has to be reduced to a principle, rather than a law, because entropy is at times reversible. This book has already dealt with the absurdity that a reversal of entropy is the same thing as a reversal in time. 3) *Darwinian evolution* indicates that life arose in a primordial soup of organic compounds as a random chemical fluctuation. Once life arose, random fluctuations in DNA chemistry created mutations, and the mutations that were adaptive enabled the organism to survive and reproduce. Following up on the *randomness producing order by accident idea* is the theory of behaviorism in psychology. Behaviorism indicates that animal engage in random behavior, and if a behavior happens to bring a reward (satisfaction of a survival drive), that behavior is more likely to be repeated and remembered. In all these theories we see order coming out of chaos by chance.

Order Theories: There are two concepts that indicate where order and non-randomness come from. One is that this order is immanent or innate in natural law, and the other concept is that order comes from a transcendental source such as a superordinate being that is separate from creation itself.

Regarding Darwinian evolution, scientists of the Intelligent Design persuasion would not necessarily disagree with the evidence that shows that complex life has evolved from simpler forms of life. Where the contention lies is whether the mutations or changes that lead to adaptation and speciation are purely random. There is a growing body of evidence that suggests that these changes are non-random – in a word, they may be teleological. Some scientists take issue with Darwinism in micro evolution where random interaction of particles seems insufficient to explain how life emerged from the inanimate world to begin with.

Dean Kenyon (2002), biochemist, originally a Darwinist, thought he had figured out how complex proteins self-assemble from amino acids based on the laws of chemistry where positive ions are attracted to negative ions. However, he found that if you mix up a soup of amino acids, they do not self-assemble into the complex proteins that life depends upon. Kenyon found that "assembly is required" by pre-existing information in the DNA. This information is required to assemble the amino acids in the correct sequence to make life-giving proteins – they cannot be put together in any random combination. Of course, this finding begs the question of where the information in DNA came from originally. Is the information inherent in nature or does it come from some superordinate being, is it personal and conscious, or is it impersonal and unconscious? These are questions that are perhaps beyond the scientific domain, but Kenyon has injected teleology back into evolution and now rejects the idea that a random dance of atoms could have created life.

Another biochemist, Michael Behe (1996), has found evidence for Intelligent Design. While Intelligent Design often gets mixed up with Creation Science (young earth, each species created separately) which has been promulgated by religious non-professionals, Dr. Michael Behe makes a clear distinction between the two and accepts most of the tenets of Darwinism – except for the randomness of mutations. Behe has studied the flagellum motor of certain bacteria and has found that the motor has the same basic parts as an outboard motor on a boat. His contention is that for the whole motor to work, all the parts must be assembled in a certain order and if one part is missing or assembled incorrectly, the motor will not work. He calls this principle *irreducible complexity*. What he means by this term is that for a complex bio-machine like this motor to work, it cannot be reduced to individual parts evolving one at a time. In other words, the whole is greater than the sum of the parts – the motor must all come together at once, not one part at a time. This concept flies in the face of Darwinian Theory which indicates that evolution comes about by small, gradual changes rather than whole systems like this flagellum motor. For example, if the bacteria received a mutation for a tail, the tail would be useless and probably not passed on to the next generation since it would not be functional and convey no survival/reproductive advantage without all the other parts such as a rotor and stator. For it to convey a survival advantage over bacteria with no tail, the whole motor would have to come into existence all at once.

The counter to Behe's argument from the Darwinist side is known as *cooption*, that is, the parts from other structures that serve another function could be coopted or borrowed to perform a different function in another structure. The problem with this argument is that all the parts would have to be coopted at once to form a whole, functional motor. Moreover, the principle of small, gradual changes in evolution in which one part would be coopted at a time would not create a whole motor completely assembled and functional – and so, would convey no survival advantage. Scott Minick (2002), a biochemist who has worked on the flagellum motor for many years, points out that cooption is inadequate in reducing irreducible complexity. He points out that 10 of the 30 parts found in the *flagellum motor* are also found in the *needle-nose cellular pump* in some bacteria. However, he asks the crucial question of how to get the other 20 parts that are needed to make the flagellum motor functional. Not only are the extra 20 parts needed, but they have to be assembled in a very precise sequential order. To accomplish this task requires other molecular machines to manufacture the parts and assemble them (Unlocking the Mystery of Life Video).

It is interesting that each side of the debate uses the *anthropic principle* to buttress their arguments. On the randomist-materialist side of the argument, its adherents claim that the Big Bang accidentally set the unique, fine-tuned constants in the new universe without a creator being. Once these were set, human life and consciousness were inevitable. On the other side of the coin, the Intelligent Design adherents claim that the constants were fine-tuned by a creator and that evolution of human consciousness would not have evolved if s/he had not guided the process every step of the way, for example, guiding the assembly of complex proteins that are the stuff of life. Fred Hoyle, British astronomer, who discovered how heavy elements (such as carbon) that make life possible are synthesized from hydrogen and helium in stars and supernovae, invoked the Anthropic Principle to account for the abundance of life-chemicals in the universe. He made the remarkable prediction (based on the prevalence on Earth of carbon-based life forms) that there must be an undiscovered resonance in the carbon-12 nucleus which facilitates its synthesis within stars.

It was also this work that caused Hoyle, an atheist until that time, to begin to believe in the guiding hand of a god (what would later be called "intelligent design" or "fine tuning"), when he considered the statistical improbability of the large amount of carbon in the universe, carbon which makes possible carbon-based life forms such as humans (Masten 2006).

An argument for the notion that order and non-randomness is innate in nature comes from quantum mechanics. Now, on the one hand, quantum theory indicates that there is much randomness in subatomic behavior as detailed in the Heisenberg Uncertainty Principle. Not only is the complementary states of particles unknowable because measurement devices interfere with particles, but unpredictability and randomness is built into the nature of the particles themselves. On the other hand, when it comes to quantum entanglement and delayed choice phenomena, it appears that randomness is replaced by orderly, predictable particle behavior. To wit, entangled particles demonstrate that complementary spin and polarization can be "known" at the same time at least by the particles. If the polarization of one entangled particle is changed, the other particle assumes the opposite, complementary polarization simultaneously (faster than light). This finding seems to remove the uncertainty of a particle's state. Furthermore, in delayed choice experiments designed by Wheeler, if the experimenter changes a condition after the particle is in flight, the particle seems to "know" about the change instantaneously and adjust its behavior instantly to appear as a particle or wave as the situation requires. Perhaps not many physicists would agree with Bohm's animistic view that all particles are conscious and intelligent, but most might agree that information is exchanged between particles so that their behavior is non-random and correlated. Again, in this case, one might argue that particle behavior, upon which mutations depend, might can be non-random and perhaps give an organism the mutations it needs to adapt and survive. Some have argued that the Cambrian explosion of life, which involved "punctuated equilibrium" or perhaps "punctuated evolution," is an example of non-random mutations.

Again, the argument goes on and on with each tribe trying to plug the loopholes in incomplete theories. Yours truly, as stated previously, believes that this issue cannot be resolved intellectually and scientifically despite all the technology of modern science. As Michio Kaku has said, the issue of Intelligent Design is "undecidable." One's belief in an intelligent designer still boils down to a matter of faith, as it always has. And so does the belief in materialist evolution – even here, there is a leap of faith or inference.

Hawking's Contribution to the "Nothing Theory" (Hawking Radiation)

Hawking's theory of radiation near a Black Hole supposedly links the very small (quantum particles) with the very large (the cosmos governed by relativity). Hawking radiation attempts to explain how a Black Hole dissolves and becomes nothing, losing all information associated with it. The basic idea of a Black Hole is that a star (or aggregate of stars) has been collapsed by gravity overcoming repulsive thermal energy, and that its gravity is so strong that that the escape velocity is c+ (greater than the speed of light). Of course, if the escape velocity is c+, then not even light (or any form of electro-magnetic radiation) can escape. Hence, one might think that a black hole would be eternal since nothing can escape its grasp. However Hawking shows that the Black Hole does lose mass and therefore will eventually dissolve into nothingness. It happens like this:

1) The quantum force in empty space is constantly hatching pairs of virtual particles that come into existence by borrowing energy from the vacuum and disappearing very quickly thus paying the energy back to the void. One particle of the pair is matter such as an electron and the other is anti-matter such as a positron. (Carr 2008: BBC Documentary).

2) Normally when matter and anti-matter meet, they are said to annihilate each other. However, "annihilation" is a misnomer because they do not go out of existence, they convert each other's mass to energy or radiation (photons).

3) When a pair of virtual particles pop into existence near a Black Hole, the anti-matter is sucked into the black hole thus reducing the black hole's mass by converting an equivalent amount of matter in the black hole to energy. But, of course, the conversion of mass to energy does not really lessen the mattergy of the black hole since energy has mass also. Unless the energy or radiation escapes, then the mass-energy of the Black Hole has remained the same. But, the concept of a black hole is that not even light and radiation can escape.

4) On the other hand, the matter particle (the electron in this case) that was spawned with the anti-matter is able to escape the gravity of the Black Hole. Now, since the electron was not annihilated by the positron, it is able to make a transformation from being a virtual particle into being a real particle (thus violating the conservation law). And the fact that it escapes the conversion to radiation by its anti-matter partner in no way lessens the mass of the Black Hole.

5) According to some theorists, as the mass of the Black Hole is reduced, it loses some of its gravity, and, as a consequence, thermal energy (which is repulsive) causes it to explode (similar to the Big Bang explosion on a small scale). Thus, the conservation of mass and energy law is preserved.

6) From this explosion, a new star may be born and thus the matter-energy of the universe is recycled and conserved as the law of conservation would dictate (BBC Documentary 2008: Hawking Radiation).

The implications of the theory are contradictory. On the one hand the disappearance of a Black Hole would seem to support the "nothingness" theory that all mass will return to nothing. On the other hand, it seems to support the idea of an eternal universe since the matter-energy (mattergy, if you will) is recycled at least on the galactic scale. Of course, Krauss might argue that eventually dark energy will rip the galaxies apart and return them to the nothingness of space. But even if matter is converted to radiation, there will still be "something" in the universe. Krauss would probably cling to the absurd notion that since radiation is made of light, and light is made of photons and photons are *massless,* then when the universe is reduced to photons there is no more universe. Therefore, he can equate *massless* photons with nothingness. Of course, the notion of massless particles reduces to absurdity because no particle of matter or energy can be massless – otherwise Einstein was wrong about the interchangeability of mass and energy ($E=MC^2$) and energy is said to have gravity (mass) as well as matter.

There are several semantic problems with Hawking Radiation similar to other theories of Modern Physics.
1) **Matter and anti-matter do not destroy each other in the sense of going out of existence.** They convert each other to energy, and according to Einstein, matter and energy are interchangeable and each can convert to the other. Thus, the anti-matter does not disappear in the Black Hole, it merely converts to energy when it encounters matter and it continues to have mass. Then the only way for the Black Hole to lose mass or energy is for radiation to escape. However, the basic concept of a Black Hole is that any mass or energy (and energy has mass) would have to exceed the speed of light to escape and not even light energy can exceed that speed according to relativity which Hawking believes in. The only way to circumvent this law is to assume that energy has zero mass and therefore can exceed the speed limit of light. But then the photon, a particle of light energy, is said to have zero mass and since it cannot exceed its own speed limit, light cannot escape the Black Hole. Hence the theory is full of entangled contradictions. It is hard to believe that nature is so contradictory and confused.
2) **Converting energy back to matter:** To further illustrate the point that matter and anti-matter do not annihilate each other but transform each other to energy, Breit and Wheeler in 1934 suggested that it would be possible to turn light into matter by smashing together two particles of light (photons), to create an electron and a positron. This 'photon-photon collider', which was predicted to convert light directly into an electron and positron pair has now been tested with positive results (Pike et al. 2014). This research further demonstrates that photons are not massless and that finer particulate matter (photons) can be converted to larger particulate matter (electrons and positrons). Furthermore, it shows that the matter-energy distinction is a false duality. Energy is simply a finer, faster, more energetic form of matter which has mass or weight as demonstrated previously. Quantum mechanics and Planck's constant indicated that energy is particulate and therefore a form of fine matter. Of course, if high speed photons in the form of gamma rays can be converted to an electron-positron pair, the pair would transform back to radiation (photons) unless they could be separated after forming.
3) Moreover, since the matter particle doesn't get destroyed by anti-matter near a Black Hole, the creation of matter out of nothing again violates the law of conservation of mass and energy. It would seem that more and more matter would be created thus increasing the imbalance between matter and anti-matter. Perhaps Hawking would argue that this is how matter came to be so much more prevalent than anti-matter especially if the Big Bang operated anything like a Black Hole.
4) Furthermore, Hawking radiation is unnecessary to explain the radiation surrounding a black hole. As gas and dust are sucked into the black hole, it speeds up and causes collisions of particles which generates friction manifesting as heat and light. That is the reason a black hole appears white, rather than black, because it is surrounded by high energy particles around the black hole and this is where the radiation comes from. Actually, a highly condensed massive body could create the same phenomena without becoming a black hole.

Allow me to clear up some of the semantic confusion among the concepts of positive/negative charge, positive/negative mass, and matter/anti-matter. In physics, mass is distinguished from matter in that mass is defined as inertia or the resistance to acceleration. Thus, negative mass would be repulsive to other masses unlike positive mass which exerts an attractive force on other mass (i.e., gravity). Of course, the separation of mass from matter is pure mathematical abstraction since the only thing that is known to have inertia and is resistant to acceleration is matter, so mass can be separated from matter only in the mind. However, when physicists speak of negative mass, they are separating it from matter. They do not mean that there is matter that is less than zero matter (that would be absurd), they mean that negative mass manifests negative gravity, that is, the object with negative mass is repelled by gravity not attracted by it, or more properly, repelled by objects, rather than being attracted to them. Of course, we know that objects of the same electrical charge repel each other, but they are speaking of repulsive matter than is not necessarily electrically charged. I don't believe any such matter has been observed and that negative mass is a mathematical construct. The following is a chart that distinguishes among the three concept pairs mentioned above.

From the chart below, we see that matter or anti-matter can have either charge, but is said to have positive matter or mass (attracted by gravity), but not negative mass (repulsed by gravity). Anti-matter does not necessarily have negative charge; however, whatever charge matter has, its anti-matter particle has to have the opposite charge (e.g., electron/positron – proton/anti-proton).

Relationships among Positive/Negative Charge - Positive/Negative Mass - Matter/Anti-Matter		
Electric Charge	Mass (Defined as Inertia)	
	Positive	Negative
Positive	Matter or Anti-matter	Neither matter nor anti-matter
Negative	Matter or Anti-matter	Matter or Anti-matter

What's the matter with anti-matter?

Does anti-matter really exist in the physical world or is it a mathematical construct like so many things in physics? It certainly began as a mathematical construct. When Paul Dirac was trying to integrate Special Relativity with Quantum Mechanics, he discovered that the math indicates that there should be positive matter and negative matter. A simplified illustration of Dirac's equation is $x^2 = 4$. This equation has two possible answers – the answer could be either +2 or -2. Thus, the math implied that there is positive matter and negative matter, but since negative matter (less than no matter) didn't make any sense, he reasoned that there must be anti-matter that is exactly like its matter twin except for having opposite charge, for example, the electron as a negatively-charged matter particle has a twin positively-charged, anti-matter particle called a positron. Since they are of opposite electrical charge, matter and anti-matter attract each other and annihilate one another (or, more accurately, convert each other to energy or radiation). Thus, matter and anti-matter cannot coexist in a material state (CERN Document Server, http://home.cern/topics/antimatter).

Negative Matter having Negative Mass: Real or Unreal?

Negative mass is the hypothetical idea that matter can exist with mass of the opposite gravitation to the ordinary stuff. Instead of 2 pounds, a lump of negative mass would weigh -2 pounds. That is, negative matter would have a negative weight or mass because it would be repulsive rather than attractive to positive matter. To measure it, one would have to have a scale that measured the downward pull of gravity and an upward push of repulsion to gravity. Instead of pushing down on a scale giving a reading of 2 kg, negative matter having negative mass would measure an upward repulsion of -2kg (a force in the opposite direction). So, the idea of matter and anti-matter involves opposite electric charge of otherwise identical particles; while positive matter and negative matter involves an opposite reaction to gravity.

The idea of negative matter/mass is based on mathematical speculation rather than observation, but there has been much mathematical analyses to determine its properties. Specifically, physicists have debated whether negative mass would violate various laws of the universe, such as the conservation of energy or momentum and therefore cannot exist in the known laws of physics. These analyses, which in the opinion of this author is little more than speculation piled upon speculation, suggest that although the interaction of positive and negative mass produces counterintuitive behavior, it does not violate conservation laws.

Cosmologists have also examined the effect that negative mass would have on the structure of spacetime. They generally conclude that negative matter cannot exist because it breaks one of the essential assumptions behind Einstein's theory of general relativity.

However, Saoussen Mbarek and Manu Paranjape at the Université de Montréal in Canada say they've found a solution to Einstein's theory of general relativity that allows negative mass without breaking any essential assumptions. Their approach means that negative mass can exist in our universe, perhaps in pairs of positive and negative mass particles in the early universe (2014). Their conclusion has far-reaching ramifications. They posit that if positive and negative matter particles exist in the universe, they would form a plasma that would have important implications for the future of astronomy.

Negative mass would also have implications for space travel. Since negative mass would be anti-gravitational and anti-inertial, it could be used to resist the force of gravity and achieve relativistic speeds. It could also be used to overcome inertia so that sudden, sharp turns could be made by a spacecraft without tearing it apart. Some UFologists claim that aliens use this in their spacecrafts to travel great distances at very high speeds and make sudden sharp turns without being ripped apart.

Evidence for Negative Mass or More Hype?

According to physic.org "Washington State University physicists have created a fluid with negative mass, which is exactly what it sounds like. Push it, and unlike every physical object in the world we know, it doesn't accelerate in the direction it was pushed. It accelerates in the opposite direction it is pushed" (Sorensen 2017). In addition, if the gravity of some massive body such as the earth pulled negative matter downward, the negative matter would pull in the opposite direction (upward). This behavior of matter is rather like the "contrarian clowns" of Plains Indians fame who do everything backwards from the rest of the tribe. If this research is true, then it would seem that gravity is like electromagnetism in having opposing forces. Here is the research evidence for this laws-of-physics-defying anti-gravity.

1>The Washington State team cooled Rubidium atoms to near absolute zero where particle motion slows to a near stop. In this slowed down state, particles become like waves which synchronize and flow almost without friction.
2>Hot, high energy particles are allowed to escape like steam, further cooling the remaining particles.
3>A laser is used to trap the atoms in a small space as if they are in a virtual bowl crammed together.
4>To create negative mass, a second set of lasers is used to reverse the spin of the atoms.
5>When the trap is released, the atoms appear to repel each other spreading out from the center.

Normally, one would expect atoms to expand with higher temperatures because of the random collision of atoms as Einstein showed in Brownian motion. However, one would not expect particles to spread out in a cool temperature near zero Kelvin. Hence the conclusion is that by reversing the spin of the atoms, gravity becomes repulsive rather than attractive.

Critique of Negative Mass Experiment?
At first blush, it would seem that this kickback of atoms released from constraints by a laser could be explained in alternate ways. It sounds very much like the Classical phenomenon of Newton's Third Law in which for every action there is an opposite and equal reaction. If one forces atoms together in close proximity, there would seem to be a kickback when that force is released - rather like the kickback from firing a weapon or releasing a compressed spring. Furthermore, if I am using my car to push another car with negative mass, it would seem that the pushed car's inertia would not only resist acceleration, but its negative mass would push back in the opposite direction against my push. Thus, the push-back of negative mass would cancel my positive-mass car's efforts and we would be at a standstill - unless I could produce enough force to overcome the negative car's inertia and its push back against my push forward. In a sense, inertia, which resists acceleration, is like negative mass in the sense that it pushes back when a force is applied to it so that conservation laws are not violated. However, it seems that these experimenters are saying that there would be not only passive inertial resistance but an active push back so that conservation laws would seem to be violated unless mathematical tricks can be conjured to rescue this contradiction.

Additionally, the rubidium atoms in the experiment are cooled to near 0-Kelvin but they are not frozen to zero motion. Since there is some particle motion causing collisions, perhaps there is enough pressure to cause the kickback that is observed when the pressure holding them together is released. It is claimed that the particulate motion is synchronized to form a wave, but there would still be collisions that cause the wave to propagate.

1> Here is what physicist, Sabine Hossenfelder (2017) says about the extravagant claims of these researchers. The experimenters created a negative *effective* mass - not an actual negative mass. Negative effective mass is not an innate property of the atoms themselves but is something artificially imposed upon them. "What this "negative mass" means is that if you release the condensate from a trapping potential that holds it in place, it will first start to run apart. And then no longer run apart." This research does not, as claimed, explain black holes, dark energy which causes the expanding universe, or any other weird phenomena of the cosmos.
2> David Abergel, a physicist at the Nordic Institute for Theoretical Physics said:

When they turned off one of the trapping lasers, some of the rubidium atoms spread and pushed themselves apart. But some of them didn't spread out, and even moved in the opposite direction the physicists predicted. This throws into question whether the experiment did indeed show negative mass or whether the expansion could be caused by other factors. In the type of physics I work in, having a negative effective mass is quite routine...It's an interesting result... but it's not turning physics upside down (Mandelbaum 2017).

This experimental claim again demonstrates the tendency in Modern Physics to make extravagant claims about the results and implications of an experiment.

Dualities of Mind and Dualities in Nature
Now there are certainly dualities in nature as mentioned above. The question is whether an Hegelian synthesis is possible to unite all these dualities in nature into one unified field theory. The question also arises as to whether these things are dualities of mind or dualities in nature.

~ matter is positive vs. energy which is negative
~ positive spin of galaxies vs. negative spin of galaxies
~ Every action has an opposite and equal reaction
~ positive vs. negative charge
~ positive vs. negative mass
~ matter vs. anti-matter
~ dark matter vs. dark energy

All the above are said to balance out to zero except for dark matter which makes up 26% of the universe and dark energy which makes up 70% (the other 4% is visible matter). The other exception to zero balance is matter/anti-matter with positive matter making up .9999999...of the normal (non-exotic) matter while anti-matter makes up a tiny fraction. However, if there had been equal amounts of matter and anti-matter, there would be nothing but radiation in the universe. Thus, if the Big Bang is true, there had to be more matter created than anti-matter. The question remains, though, as to whether the idea that there is a twin anti-matter counterpart to every matter particle is a thing hatched in the dualistic mind of mathematicians or whether it exists in nature. Scientists at CERN claim to have made tiny amounts of anti-matter. Furthermore, the PET (Positron Emission Tomography) scan used in medical diagnostics is said to be based on positron/electron (matter/anti-matter) annihilations. However, we have seen that technology and mathematics do not, in themselves, prove a particular theory.

> *According to physics theory, the universe is a zero sums game except for matter/antimatter and dark matter/dark energy.*

Susskind's Refutation of Hawking Radiation and Information Loss in a Black Hole

Leonard Susskind (2008), physics professor at Stanford University, has taken issue with Hawking radiation and the loss of information in a black hole. Perhaps his long debate with Hawking could be described as a kind of Holy War to preserve the purity and sanctity of quantum physics. Actually, Susskind called it the Black Hole War in which he aims to make the world safe for Quantum Mechanics. Susskind took Hawking to task when Hawking contended that information is irretrievably lost in the dissolution of a black hole (as previously described). Susskind argues for another conservation law, namely, the conservation of information, i.e., that information is never lost in any interaction in the universe because the math tells us so. Susskind says that the information about the distinctions among elementary particles is never annihilated. For example, the information that a proton is made up of three quarks is not lost even though a proton might be broken apart and the three quarks released as independent elementary particles. Susskind creates an hypothesis equally as preposterous as the one for which he criticizes Hawking. Susskind says that the information about particles is preserved in a two-dimensional holographic ring around the black hole. Not only is such a 2-D holographic information ring preserved around a black hole, but around the whole universe as well so that no information is ever lost from the universe. Now a hologram is a three-dimensional projection that is said to be projected from a 2-D piece of film with laser light, but, of course, the film itself is three dimensional. Furthermore, the assumption behind a universal hologram of information is that the universe is finite and spherical in shape. Even though the math, properly configured, may support this incredible idea, what does logic and nature have to say about it?

1) To begin with, as pointed out previously, a particle is not annihilated by a collision with its anti-matter counterpart in a black hole – it is converted to radiation or energy (high speed photons of various frequencies). Now if matter and energy are interchangeable, then not only can matter be converted to energy, but energy can be converted back to matter – otherwise the conservation law is of no effect. That information for putting matter together would seem to be inherent in the matter itself and would not need to be stored in a hologram around a Black Hole or the universe – it would seem to be stored in the mattergy itself. Thus, no information is lost when something is sucked into an hypothetical black hole and then the black hole dissipates. To use a chemical example, if I burn a piece of coal by combining it with oxygen, the resulting carbon dioxide preserves the matter and information about each element in this molecular form and thus the information about carbon and oxygen is not lost in this chemical combination. The CO_2 can be broken apart again so that carbon and oxygen preserve their initial form and properties. Regardless of what nature says about the issue, it seems to be the consensus in the physics community that Susskind has won the argument, since his theory is now in physics texts, and Hawking has finally acknowledged defeat. However, perhaps both are wrong in this highly speculative science.

2) In actually, this debate is precipitated by the inherent contradiction between relativity (in which Hawking specializes) and quantum mechanics (in which Susskind specializes). Susskind calls this conflict between the two great pillars of modern physics a paradox. However, a paradox is an *apparent* contradiction that can be logically resolved; whereas a real contradiction cannot be resolved. I see this conflict as a real contradiction, not a paradox to be pseudo-resolved by patches and hedges designed to prop up shaky theories resting upon a foundation of quicksand. If Susskind is a true quantum physicist, then he must embrace the uncertainty principle and the

notion that quantum behavior is largely random, i.e., without lawfully-deterministic information. Quantum physicists tell us that quantum randomness is not a result of the ignorance of the physicist, but that it is inherent in the nature of things. Einstein was repulsed by this lack of lawful information in quantum theory and protested that "God does not play dice." So why is Susskind concerned with the loss of information in a Black Hole when much quantum interaction is largely random, i.e., without information to determine its ultimate form?

3) Susskind's holographic library would seem to contradict Krauss's concept of the universe coming from nothing and returning to nothing. On the face of it, if the universe truly comes to naught, then all the information that once held it together is lost forever. But Krauss engages in much double speak and equivocation in propounding slippery concepts. The matter of the universe does not cease to exist but is transformed to radiation (photons of various wavelengths); however, this radiation is so scattered by dark energy that it would be impossible to materialize again. BUT, recall that Krauss and other quantum physicists say that virtual particles can be hatched from the vacuum of space and that some of these quantum particles can metamorphose into real particles. In order to do this, the vacuum of space would have to contain information on how to create symmetrical matter-antimatter particles. So, to Krauss, the information for making another universe is contained in the vacuum of space (nothing). Indeed, Krauss argues that the vacuum has probably already created other universes from this information in the void.

Hawking Radiation: Clearing up the Confusion

There is much confusion about Hawking radiation and the evaporation of black holes. Here are two interpretations of this process that I have encountered – neither of which holds up to scientific scrutiny.

Matter-Antimatter Annihilation Interpretation

A matter and anti-matter particle are hatched from the quantum foam near the event horizon of a black hole >Rather than being attracted to each other and annihilating (converting to energy), the antimatter particle is pulled into the black hole, but somehow the matter particle escapes > the anti-matter particle annihilates a matter particle thus reducing the mass of the black hole > since the matter and antimatter annihilation coverts to gamma rays carried by photons and a photon is said to be massless, the mass of the black hole has been reduced >since the matter particle escapes the gravitational clutches of the black hole, the mass-energy conservation law has not been violated since an equal amount of mass that was created was destroyed in the black hole.

Fallacy: The main fallacy of this interpretation is that the annihilation of matter and antimatter particles means they go out of existence. Actually, they convert to energy (gamma radiation carried by photons). However, since electromagnetic energy cannot escape a black hole because of its strong gravity, the mass of the black hole remains the same since energy has an equivalent amount of mass. Therefore, there is no dissolution of the black hole.

But here is what Tomas Zato of Physics Stack Exchange (2014) says about the matter-antimatter annihilation interpretation:

*(This interpretation) confuses anti-matter and **negative matter**. Both matter and anti-matter are affected by gravity...When it's annihilated, it yields energy worth its mass. Negative matter, which is purely theoretical, would be pushed away from normal matter, reacting inversely to gravity. It would, as well, curve spacetime "up" instead of "down". Some have suggested that this could allow one to construct structures which would (permit) exceeding the speed of light.*

Gravitationally Positive Mass vs. Gravitationally Negative Mass Interpretation

Virtual mass particles are hatched from the quantum foam near the event horizon of a black hole >One of the virtual mass particles is somehow pulled into the black hole and becomes a negative mass real particle >This negative mass particle cancels a positive mass particle $[(+1) - (-1) = 0)]$ thus reducing the mass of the black hole without any ejection of matter or energy unlike the radiation that is given off with a matter-antimatter annihilation > Since the other virtual mass particle escaped the black hole, it had to become an actual positive matter particle to balance the loss of the negative matter particle so that the conservation of mass-energy law is not violated.

One would think that as in electro-magnetism, opposites would attract and likes would repel, but it is not that way in hypothetical gravitationally positive and negative mass. The following table indicates the hypothetical relationship of positive and negative mass.

Actor	Acted Upon	
	Mass+	**Mass-**
Mass+	Attracts	Attracts
Mass-	Repels	Repels

* In other words, positive mass attracts both kinds of mass
* Negative mass repels both kinds of mass.
* When equal amounts of positive and negative mass are put together, the positive mass attracts but the negative mass repels, so that the resulting gravity is net zero.
* When negative mass is put with negative mass, each repels the other perhaps providing double acceleration.

Fallacies

1> The main fallacy is what I call the "zero fallacy" which is the misuse of the meaning of zero in modern physics. To begin with, there can be no negative mass particle in the sense of a particle that has less than no mass. Just as there can be no massless particle (zero mass), there can be no particle with less-than-nothing mass. Some physicists get around this untenable idea by saying that a negative mass particle is negative in the sense that it exerts repulsive gravity rather than attractive gravity. Hence combining a positive mass particle with a negative particle of equal mass yields zero. But the zero refers to no net gravity, not "non-existence". It does not mean that the positive mass particle and the negative mass particle obliterate each other and cease to exist in any form. They continue to exist but cancel each other's gravity.

2>A virtual negative particle is supposed to repel a virtual positive particle; however, since a black hole is made of mostly positive mass, the greater strength of the black hole would pull it in. But virtual particles are supposed to be massless (although there is equivocation on this) and therefore would not respond to gravity (again there is equivocation on this since photons which are allegedly massless react to gravity). If massless particles do not respond to gravity, then the black hole would not pull it in.

3>If the virtual negative particle is *massless*, then it could not annihilate a real *mass* particle. Only a real negative mass particle could cancel a real positive mass particle's gravity. Thus, somehow the virtual negative mass particle must convert to a positive mass particle in the black hole.

4>The final result of Hawking radiation over time would be an aggregation of positive mass and negative mass particles which would balance out the gravity of the black hole to net zero. Then there would be nothing to hold the particles together, and the particles would be driven apart by thermal radiation or other force and the black hole would dissolve into a sea of scattered particles. However, the hole would not evaporate into nothingness as originally proposed.

5>What seems to be confused here is antimatter and negative matter. If antimatter reduces matter to radiation (gamma rays carried by photons), then one could argue that since photons are supposed to be massless, that mass has been destroyed resulting in the death of a black hole. That idea is refutable, but it is better than the idea of negative mass destroying positive mass because negative matter doesn't annihilate positive matter – it merely repels it.

6>As Susskind points out, if positive and negative mass particles truly obliterate each other out of existence, there would be a loss of information from the universe which Susskind says is impossible. What Susskind proposes to prevent this loss is equally speculative. He says the information would be preserved in some kind of two-dimensional holographic ring around the black hole – perhaps carried by the particles that didn't get sucked in. Of course, if the black hole dissipates, the information would seem to be lost.

7>There is much equivocation about whether anything can escape a black hole including electromagnetic radiation such as gamma rays. The original assertion was that nothing, not even light, could escape the clutches of the strongest gravity in the universe. Then Hawking introduced his idea of mass-killing radiation which means that the black hole evaporates without anything being emitted from it. Instead, according to Hawking, the death of the hole is because of mass-killing, virtual negative-gravity particles. In this view, the radiation seen around a black hole is caused by the bubbling up of particles around the hole out of the quantum foam and perhaps by acceleration of particles being pulled into the hole. Others, however, argue that radiation does indeed escape from inside the black hole thus contributing to its eventual death. Since Hawking's idea of gravitationally positive and negative particles obliterating each other without a trace of radiation is illogical, it would seem to this author, that if black holes die, there has to be some leakage of mass and energy. It's simple arithmetic if the mass-energy conservation law is true. You can't get by with destroying matter even if you postulate that a virtual particle turns real to compensate for the loss. What is the law of nature that says that a virtual particle turns real when its twin virtual particle dies? One might call on entanglement of particles, but that is a stretch.

Gravity: A Push or Pull, Attractive or Repulsive?

A study released in March of 2017 suggests that two black holes have merged because of their two host galaxies colliding. Before the merger, the two black holes began swirling around each other, and this energetic action produced gravitational waves. As the two hefty objects continued to radiate away gravitational energy, they moved closer to each other over time. If the black holes do not have the same mass and rotation rate, they emit gravitational waves more strongly in one direction than another. The black holes finally merged, forming one giant black hole. The energy emitted by the merger propelled the black hole away from the center in the opposite direction of the strongest gravitational waves (Hille, NASA.gov 2017). This effect sounds rather like Newton's third law - for every action, there is an equal and opposite reaction. It would be similar to the kick back (recoil) one feels when firing a gun.

Wall in Science.com states:

The monster black hole has already zoomed 35,000 light-years away from its galaxy's center, farther than Earth and its sun are from the core of our own Milky Way. And the behemoth is currently traveling outward at 4.7 million mph (7.6 million km/h)—fast enough for the black hole to escape its galaxy completely in 20 million years (2017).

This speculation suggests that gravity waves can be repulsive rather than attractive – in other words gravity pushes as well as pulls. In this study mentioned above, gravity is seen as a push that dislodged a black hole from its host galaxy. In this book, we have already looked at the hypothesis of *negative mass* which is supposedly repulsive rather than attractive. As per usual, no such negative matter has ever been observed, and it is a mathematical projection based on fantasy. Also, we are told by Einstein that gravity is not a force at all, but a curvature of spacetime that compels objects in motion to move toward each other; therefore gravity is seen as a push rather than a pull in General Relativity. Thus, in the view of relativity, when jumping out of a plane, the sky diver is not pulled by the earth but pushed down by the curvature of space caused by the earth's mass. If gravity is not a force, as Einstein has said, then it seems that there would be no need for waves to carry the force (similar to the kind we see with electro-magnetism). Nonetheless, Einstein hypothesized that such waves exist as a disturbance in the spacetime medium, and some physicists claim that LIGO scientists (discussed previously) have detected such waves created by the collision of two black holes.

Asked whether gravity is a push or pull, the writers at *Physics Forums* give the usual double-speak and indicate that such a question is a meaningless one and that somehow in the vagaries of relativity, gravity can be either or neither or both a push and a pull. B. Crowell, staff emeritus and science editor for *Physics Forums*, indicates that in "fields", it is not possible to say what a push or pull is:

It makes sense to define repulsion and attraction in a Newtonian universe where everything happens through instantaneous action at a distance. Once you start talking about fields, it is no longer possible to classify everything as repulsive or attractive. For example, when an electromagnetic wave passes through a region containing a charged particle, it doesn't make sense to ask whether the particle is attracted or repelled. In GR, the closest thing we have to a meaningful way of saying gravity isn't repulsive is the various energy conditions. E.g., there's the strong energy condition, which is violated by the cosmological constant. Gravitational waves don't violate any energy conditions (Physics Forums).

Whether it makes sense or not, physicists do indeed talk about gravity as being either a push or pull. Kaku says that General Relativity indicates that gravity is a push caused by spacetime curvature and that this push is what makes the sky diver fall earthward. In contradiction to Einstein, Big Bangers talk about "spagettification" that would occur if some unfortunate soul fell into a black hole. Bangers tell us that tidal forces would be so strong that for a person falling in feet first, her feet would be stretched out before her head and the rest of the body are stretched; because the feet would be closer to the center of gravity. Thus, the person's body would resemble a string of spaghetti from such elongation. Now "spagettification" sounds like gravity is an attractive force that exerts a pull rather than a push on anyone falling into the abyss of a black hole.

If common-sense rationalism were permitted in modern physics, there would be an easy experimental way to determine whether gravity is a push or pull. If gravity is a pull, then an object (such as a human being) would be elongated or stretched when being attracted to a large gravitational mass because of tidal forces. If, on the other hand, gravity exerts a push as Einstein indicated, an object would be compressed as it moves toward a gravitational mass. Of course, according to Crowell, this kind of thinking belongs in a Newtonian, Classical universe where things make sense, but such thinking has no place in the metaphysics and mysticism of Modern Physics.

As an additional critique of a massive black hole being knocked out of its galaxy by gravitational waves, it would seem that if the galaxy had lost its nucleus of gravity (the black hole), the galaxy would begin to dissipate and scatter and lose its signature structure of a spiraling pattern of mass drawn toward a center. Additionally, why would the gravity waves, if they are repulsive, not displace the whole galaxy rather than just the black hole whose tremendous inertial mass would require an almost irresistible force to move. Of course, the black hole would have more concentrated, dense mass so maybe it could be argued the gravitational waves would have more to push on. Furthermore, the thesis of this book is that semantically it is untenable to speak of space (which is vacuum) manifesting itself in waves. Additionally, even if there is a difference in size between the two black holes circling around each other, the larger black hole would be sending out waves in all directions since it circulates and spins. Even when the holes merge, a black hole is said to continue to spin, so that the waves would be sent omnidirectional. Additionally, it would seem that if the two spinning black holes are generating repulsive gravity waves as they rotate around each other, they would repel each other rather than merge and that the larger black hole would eject the smaller black hole rather than merging with it

Furthermore, since most celestial bodies spin, it would seem that they would create repulsive forces on the satellites that orbit them. For example, the earth spins at a rate of 1000+ mph at the equator – yet the moon is attracted into an orbit around the earth. Likewise, the sun spins in a sloshing way since it is a gas or plasma, and it keeps the earth and the other planets in orbit. Perhaps the cosmologist would argue that this repulsive force is small compared to the attractive force and that is what keeps the satellite in orbit rather than falling into the larger gravitating body.

The Shape of the Universe: Flat or Soccer Ball Shaped?

According to Krauss (2012) and others, the latest evidence indicates that the shape of the universe is flat. The idea of a flat earth is absurd, but the universe within which the spherical earth resides is said to be flat. At first blush, the idea of a flat universe seems as absurd as the idea of a flat earth. Just what is meant by flat and non-flat models of the universe? As usual, physicists give little attention to the precision of language and focus on mathematics which can be no better than the language that underpins the math.

1) When we think of flat, we think of 2-D space, the stuff of plane geometry where figures are drawn on a flat piece of paper. Surely, this is not what physicists mean by "flat" because on a local level, we see that material objects exists in 3-D, and we see celestial bodies in all directions occupying three-dimensional space.
2) All large celestial bodies that we see are spheroid in shape, solar systems involve elliptical motion, and many galaxies are spiral-shaped. Wouldn't these components of the universe suggest that the whole universe is spherical or elliptical? However, our solar system is disk shaped since all the planets with their moons basically follow the ecliptic and galaxies are usually disk shaped as well. Of course, galaxies and solar systems are still three dimensional and manifest curved lines – they are not 2-dimesional disks because a disk has some thickness and is therefore 3-D.
2) What physicists really mean by "flat" is that space involves straight lines as opposed to curved lines on a universal level – not that space is 2-D as in plane geometry. But, you say, Einstein taught us that mass curves space in three dimensions so that even the path of light is curved (although in spacetime, the path of objects is supposed to be straight). However, physicists tell us that these curved lines are only in our little corner of the universe where the density of matter is high, but space is mostly empty, and out there between galactic clusters, gravity is very weak, and space is dominated by repulsive dark energy. In outer space, then, lines straighten out, and in the curved space that light travels through where there is dense matter, the curvature averages out for light's path which encounters about as much positive curvature as negative curvature.

__The answer is to be understood in terms of geometry, particularly the geometry of triangles.__

There are three major possibilities as to the geometric shape of the universe: 1) flat (the universe would be infinite), 2) closed (a closed curve as in a sphere) in which the universe would be finite, and 3) open (an open curve as in a saddle) in which the universe would be infinite. Now Euclidian geometry shows us that the three angles of a triangle will always add up to 180-degrees in flat space no matter what shape the triangle takes. However, if we form a triangle on a sphere, the angles will add up to more than 180-degrees, and if we create a triangle on a saddle, the angles will add up to less that 180-degrees. Of course, this line of reasoning is a perversion of Euclidian geometry. Greek geometry is based upon metaphysical perfection and deals in the abstract; thus, Euclidian geometry is correct to say that if you draw a triangle on a flat space using straight lines, the angles would add up to 180-degrees. A triangle drawn on a sphere or saddle is not a true triangle because it involves three dimensions and curved lines. I would call these para-triangles (perhaps convex or concave triangles or curved triangles) rather than simply "triangles" because they do not meet the criteria of a true triangle in the abstract. And, if there is such a thing as anti-gravity, a straight triangle could be drawn in space because gravity would not be curving the instrument that you are drawing the line with. Of course, Einstein taught that space is non-

Euclidian because of its curvature and its fourth dimension (time). Now we are told that space is Euclidian after all on the universal scale, meaning that on long distances, light travels in straight lines.

How do we know that space has straight lines (or more logically, that light travels in straight lines in the great expanse of space)?

By triangulation of course. WMAP (Wilkinson Microwave Anisotropy Probe) space craft was able to measure the hot spots and cold spots of the Cosmic Microwave Background radiation. WMAP received radiation from two distant points in space from a hot spot and a cold spot in the CMB to form a vast triangle. Using geometric calculations like a surveyor, they were able to figure that the angles added up close to 180-degrees thus concluding that on the large scale, the observable universe is flat (that is, space is made of straight lines as indicated by the path of electro-magnetic radiation).

Of course, since the observable universe is only part of the total universe, there might be slight curvature that WMAP and other instruments are not sensitive enough to detect. To wit, humans walking on the earth were not able to detect its curvature and believed in a flat earth for millennia. If the universe is flat (having straight lines), then it would seem that the Big Bang would appear as a sphere but would involve straight lines emanating from a center. In other words, light and matter would be flying in straight lines from that center and would not be captured in an orbit. However, physicists are given to mysticism, and the idea of a center to an explosion is too commonsensical and rational for their taste. Instead some prefer an exotic concept of the bang in which there is no center. Check out what Dr. Jagadheep D. Pandian (2015). of the Indian Institute of Space Science and Technology says about a center to the Big Bang:

The meaning of the Big Bang has been very often misunderstood. It is thought that something exploded somewhere and then the exploded part expanded to where we are currently. This is not correct. Before the Big Bang, there was no space or time. So, there is nothing "outside" the Big Bang. The Universe simply expanded from a very small volume into a huge volume, and this expansion is occurring even today. So, the place where we are right now corresponds to some place in a very small volume in the very early Universe. Hence, the Big Bang occurred EVERYWHERE in the Universe. It occurred at all places including the place where we are right now

Is the universe on a grand scale really that different from our local area and totally incomprehensible to the rational mind? Let's examine the logic of Dr. Pandian. "The Universe simply expanded from a very small volume into a huge volume" seems to contradict the statement: "the Big Bang occurred everywhere at the same time." If the Big Bang occurred everywhere at the same time, then what is the need for expansion since the bang is already there and everywhere? The expansion would seem redundant if space was banging everywhere simultaneously. This statement also contradicts what we have been taught, namely that the Big Bang started from a ***singularity*** in spacetime (one point in space, not everywhere or every point in space). How can the idea of one singularity possibly jive with infinite singularities all over space? Moreover, how would Dr. Pandian explain the tremendous heat of the Big Bang if there were no explosion? And why has the universe cooled as it expanded if there was no center from which heat radiated. Since the Big Bang was not observed, I would say this idea is little more than mathematical metaphysics and even mysticism. It is the "dead and alive cat" line of illogic which reduces to absurdity and should have no place in scientific realism.

As per usual, someone else has come up with a different interpretation of the CMB.

Sean Markey (2003).informs us that the shape of flat-universe models imperfectly fits the mathematical proofs derived from WMAP. In other words, taking the density fluctuations in cosmic background radiation recorded by WMAP, the math adds up if the universe is finite and shaped like a dodecahedron (a soccer ball). Thus, in this interpretation, the universe is finite.

Seeing the Big Bang

If we looked out far enough, we would see the Big Bang itself. In principle this is not impossible, but in practice, between us and that early time lies a wall. (Krause 2012: p. 42).

Allow me to address some of the other fantastic derivations from GR, SR and quantum mechanics that Krauss and others have applied to cosmology. One that particularly struck me was Krauss's contention that were it not for the walls of plasma blocking our vision by absorbing light, we could see back in time to the Big Bang. Now, since the Big Bang theoretically occurred some 13.7 billion years ago, it would seem at first blush that the event has already occurred and vanished into history. It would be rather like pointing my telescope toward Pennsylvania and seeing the whole Battle of Gettysburg replayed some 150 years after the light carrying those images of the battle have long since disappeared into space and no doubt scattered. It would seem that a time machine, if there were such a thing, might better help me see this terrible battle. But, never fear, the mathematics of modern physics allows all manner of sleight of hand manipulation of time. There are three scenarios for viewing or not viewing this seminal event of the Big Bang. 1) If the galaxies are truly being carried by space expansion at the exact speed of light, then we would perpetually see a still image of the Big Bang since we would be traveling at the same speed as the light that is carrying the image of the primordial bang of so long ago. 2) If we are being carried by space that is expanding faster than lightspeed, then we would not see the bang because we would be outrunning the light that is carrying its image. 3) If we are traveling slower than light, then the light carrying the image of the bang would have hit us and passed us by long ago. Perhaps Krauss means that we might see the echo of the Big Bang in the CMB, but that is the after-effect, not the bang itself. Furthermore, Krauss argues that the universe has not always expanded faster than light, and at one time, before acceleration, it was traveling much slower than light. If our Milky Way average speed is less than light, then the light carrying the image of the bang has passed us by and headed into deeper space. But wait, if the Big Bang expansion is creating new space, light would have no space to travel in if it passed us by. So, it seems that the walls of plasma are not the only things that prevent us from seeing the origin of our universe

Using Michio Kaku's bubble picture of the universe which would appear as a sphere in a metaphorical God's eye view, the only way we could see the Big Bang is if we had been traveling at the speed of light since the matter that makes up our Milky Way galaxy left the Big Bang. Since most cosmologists say the universe is flat (not two-dimensional but involving straight lines on the large scale), then the following graphic represents the current view and shows the impossibility of seeing the Big Bang.

Seeing the Big Bang

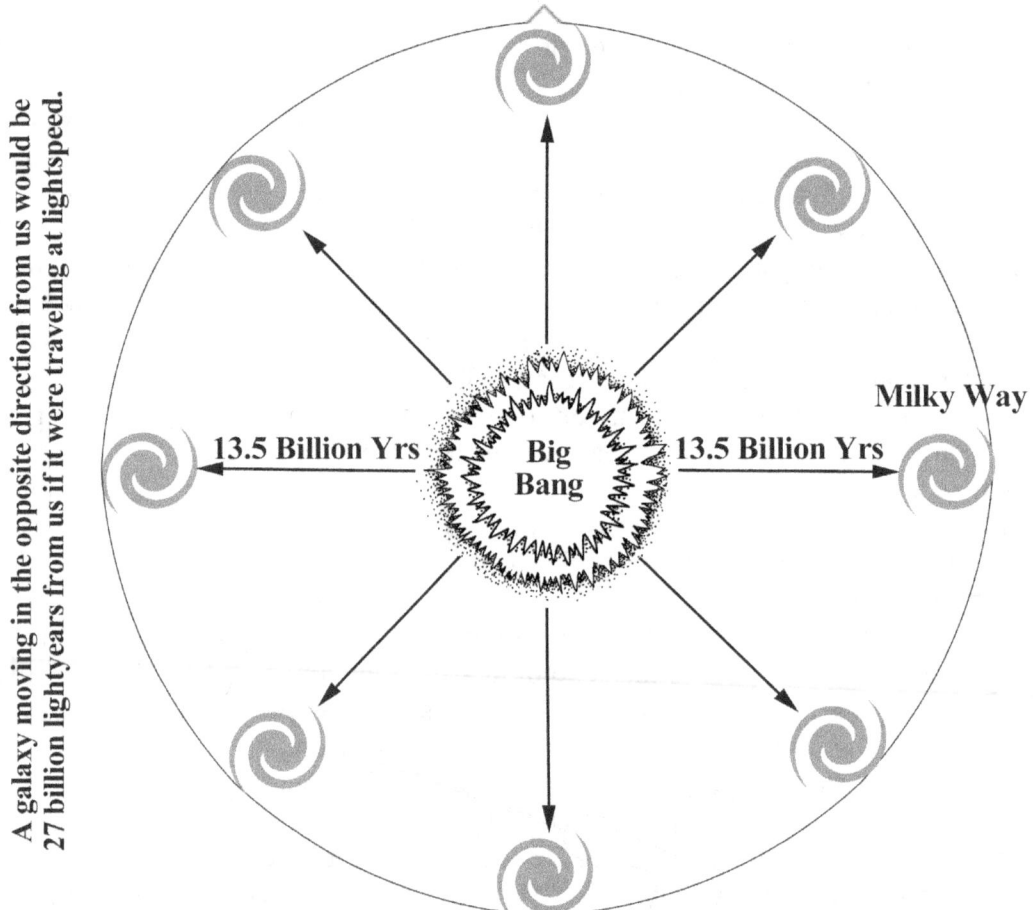

The only way to see the Big Bang is if our galaxy had been traveling at light speed for 13.5 billion years so that we would be 13.5 billion lightyears from the bang.; therefore the light carrying the image of the Big Bang would be traveling with us. If we were moving slower than lightspeed, the light carrying the image of the Big Bang would have passed us already.

There are other problems with glimpsing the genesis of the universe. According to relativity, as something approaches the speed of light, space contracts as time slows down to maintain the constancy of light speed. So, if space is expanding faster than light, then time is running backward, and space would have to expand in a negative direction (so you would get negative space, i.e., less than nothing). But, modern physics has a hedge for that dilemma too. Kaku says that space can expand faster than light because space is nothing, and nothing can go faster than light. Krauss echoes this koan, but admits it sounds a bit fishy. It's the same argument that is given for the photon being massless – because anything with mass cannot travel at light speed since that superluminous object would acquire infinite mass and that would require infinite energy. Of course, the photon has mass - if energy is truly interchangeable with mass as indicated in $E=MC^2$.

COSMIC SENSE, COMMON SENSE...   Seeing the Big Bang

So now modern physics is rescued from the contradiction that space is going faster than light because physicists conveniently invoke the concept of Special Relativity that space is nothing (massless). However, SR contradicts General Relativity which indicates that space or spacetime is something - a medium which is warped by mass and can carry gravity waves. Quantum mechanics says space is nothing, out of which arises something (a quantum foam of particles). But then we have another problem. If space is expanding and carrying the galaxies with it faster than light, then galaxies (which are certainly massive) are also traveling faster than light – thus violating the speed limit for mass. Krauss hedges on this problem by saying that the galaxies are at rest in space, and thus they are not traveling faster than light - it is the space that is carrying them which is traveling faster than light. But if something is being carried faster than light, is it not traveling faster than light also? If my car is carrying me at 70 mph, am I not traveling 70 mph also? Somehow all these fixes for relativity just don't fix it. There are so many glaring contradictions, I don't see how the theory could ever have become the established theory in physics.

Relativity, Light and Expansion Problems

Although spatial expansion is purportedly based on General Relativity, expansion conflicts with Special Relativity and the constancy of light speed in significant ways. As noted above, space expansion plays funny tricks with the constancy of light speed. Now the hedge is that light speed is not added to space speed because light, like matter, is just being carried along by space. However, this is an epicycle-type hedge that cannot stand up to logical scrutiny. An apt analogy would be the speed of sound which is said to be constant at standard temperature and pressure in still air. However, if wind is carrying the sound, the speed of the wind is added to the speed of sound like a swimmer who is swimming with a current. Likewise, if space is carrying light, then space speed would be added to light speed violating Einstein's dictum that nothing can exceed the speed limit of light, including light itself. Here are more glitches in this theory.

1>If space were expanding at c, then light is going 2c if traveling in the same direction as expansion, Light going in the opposite direction from the point of Big Bang expansion would have a speed of 4c relative to the light beam traveling in the opposite direction.

2>Thus we would never see galaxies moving away from us even if they are receding at c. If they are exceeding c, then we certainly would never see them.

3>If, as the Expansionists argue, light, like matter, is just being carried along by spatial expansion and space speed is not added to light, then light would be traveling at zero speed (standing still) relative to space since space would be traveling at light speed. Light would be treading water, rather than swimming with the current. If this were so, light would never travel between galaxies because the galaxies are allegedly going at the same speed since they, too, are being carried by space at light speed or greater. So, we see that the speed of space expansion has to be added to the speed of light. Moreover, if light speed remained constant regardless of spatial expansion, then light traveling opposite the expansion would be standing still like someone swimming against a current at the same speed as the current.

4>If stretching and shrinking space affects the speed of light and its frequency, then it would seem that light would have different speeds and frequencies in intergalactic space as opposed to intra-galactic space. Since we are told that the expansion is occurring in intergalactic space rather than intra-galactic space, light would be most stretched out there depending on the direction of expansion. On the other hand, we are told that space may be shrinking in intra-galactic space due to the dominating influence of gravity and dark matter at that level. For example, the Andromeda galaxy is said to be moving toward the Milky Way and will one day collide with the Milky Way. Of course, the rational interpretation is not that space is shrinking between the Milky Way and Andromeda, but that they are moving toward each other in space.

5>Since science fiction seems to lead the way in physics theory, we might need to invoke the Star Trek concept of subspace to explain all this. Perhaps as space expands, there is a subspace strata underneath space that was already there, and which remains constant and does not expand. Then perhaps there is a superspace also. Even so, light traveling in the space that is expanding is violating its own limit.

Black Holes: Densest Mass in Universe or Nothing?

Tryon knew that the total energy of a black hole is zero, because of the trade-off between mass energy and gravitational energy (Gribbin 2009).

This quote from Gribbin's book plays into the "zero fallacy" that we have encountered before. The illogic behind this assertion is that since the mass-energy of the black hole is equal to the gravitational energy that holds it together, the algebraic sum yields zero net energy. Yet, we are told that the black hole represents the heaviest concentration of mass anywhere in the universe. Since mass is equivalent to energy (or can be converted to energy) as expressed in $E=MC^2$, we know that the black hole contains an enormous amount of mass-energy compressed into a very small space. Perhaps we should call this massive energy "potential energy" since it is confined within the event horizon of the black hole, but it is energy nonetheless. One view of this is that gravity holds particles so tightly that there is no particle motion; therefore, no radiation in the form of heat. If this is true, then the black hole would be zero Kelvin since at 0 degrees K all molecular motion theoretically stops. There is much equivocation about whether black holes emit radiation, but there is surely radiation surrounding a black hole. If, as some argue, some of the radiation comes from inside he black hole, then the forces are not in balance and do not equal zero. In this view, as the black hole loses mass-energy, gravity becomes weaker and the expansive energy of heat and gamma radiation becomes stronger than the gravity of the black hole thus causing it to explode. Out of this explosion, new stars may arise, Phoenix-like, from the ashes. Furthermore, the fallacy is that since the energy of a black hole is approximately net zero since expansive energy of radiation is nearly equal to gravitational energy, the energy of the black hole is zero and does not exist.

Other Fallacies Regarding Black Holes

There is much speculation about black holes, and some of it defies all logic. Of course, the yogi logic of mysticism can be invoked to cover these fallacies. Here are a few anomalies that I see.

1>**The black hole represents the densest matter in the universe…yet it might be possible for a probe to pass through it to the other side** (Kaku 2011). The problem again seems to be in *semantics*. To begin with, the term black *hole* is a misnomer. A hole is an empty space surrounded by matter. A hole in a wooden wall is a space where the wood is not present and can allow things to pass through the wall. A black hole may resemble a hole because it sucks matter and energy into it, but it is the opposite of a hole in the sense that it is filled with the densest matter known to astronomers – it is not a vacuum that allows things to pass through. Hence if light cannot get out of a black hole, it is unlikely that a probe with camera could get through it. In defense of the idea of something passing through a black hole, it is said that there are two kinds of black holes – spinning and non-spinning. The images that we see of a non-spinning black hole is a black solid sphere of extremely high density which is the remnant of a collapsed large star. It is unlikely that any macro mass could get through this dense mass – especially if light and electro-magnetic radiation cannot escape. I suppose this concept of a spinning black hole is that the hole is like a whirlpool (or suck hole) in which water swirls in a funnel shape or perhaps like the eye of a hurricane where there is low pressure and low density of air. However, all the visual renditions I have seen show the funnel coming to a point or singularity. Other conceptions of a black hole include a tunnel in space time through which things pass into another universe. If there is a singularity or near singularity, that would pose a problem for a massive object to get through. Still, with a spinning black hole, if light cannot escape, it is unlikely that a macro object could pass through it.

2> **Nothing can escape the clutches of a black hole**. Yet, we are told that jets of gamma rays and gas can spew out of a black hole because of the extreme energy created by compression. This ejection of luminous gas is sometimes called a white hole and sometimes a quasar where a new galaxy is forming with a black hole in the center. Now it may be argued that when gas is being pulled into a black hole before it crosses the event horizon, it lights up because of high speeds and particle collisions that occur. However, some theorists are claiming that gas and gamma rays actually come through the black hole (inside the event horizon). Hence an amendment should be made to the effect that it is difficult, but not impossible, for any mass or energy to escape a black hole.

3> **The black hole compresses to a singularity in the center**. While there is some equivocation on what a singularity is, it was originally conceived as a geometric point which has no dimensions and therefore occupies no space. The idea of matter, no matter how compressed, not occupying three-dimensional space is surely an absurdity. Now, some cosmologists are saying that the singularity does have dimensions - albeit very small. Sean Carroll (2004) says that at the center of a black hole, as described by general relativity, lies a gravitational singularity, a region where the spacetime curvature becomes infinite (p. 205). I would say that if Carroll is correct about a singularity, the singularity comes to a zero point, not an infinity. Carrol goes on to say that for a non-rotating black hole, this region takes the shape of a single point and for a rotating black hole, it is smeared out to form a ring singularity that lies in the plane of rotation (pp. 264–265). In both cases, the singularity has zero volume. Zero volume with infinite density sounds like a mathematical artifact to yours truly.

Alternatives to the Big Bang: Steady State and Quasi-Steady State

Back to the quintessential question: Is the universe eternal or did it have a beginning, and will it have an end? Judeo-Christian theology says that the physical universe had a definite beginning in time, but Buddhists say that the universe is eternal. Steady State theory championed by Hoyle, later modified to form the Quasi-Steady State theory, argues that the universe is eternal and Big Bangers, of course, argue that the universe we know had a definite beginning in time, and they can describe the hatching of the universe down to the billionth of a second. Steady State theory is based on a concept that we were taught in biology of nature's massive recycling program that permits no waste of organic material. The processes of bio recycling are called anabolism and catabolism, and they bear out the conservation of mass and energy in the biological realm. Catabolism is the breaking down of organic matter into simpler compounds, and anabolism involves taking those simpler compounds and building them up into larger compounds to be used by a living organism. When an animal or plant dies, its body breaks down into simpler compounds which are taken up by another plant or animal whose system synthesizes these simpler compounds back into more complex compounds such as proteins needed by the organism to survive and function. Of course, DNA and RNA provide the information and machinery for this synthesis of amino acids into long protein chains.

So it is with the Steady State cosmology. When a star or galaxy dies because it has burned its matter converting it to radiation, the resulting radiation and particles are reassembled to create new stars and galaxies using the DNA of natural law (interchangeability of mass and energy), implying that this process could go on endlessly if the law of conservation of mass-energy is true. On the other hand, Big Bangers initially claimed that the universe sprang out of a singularity (nothingness) and will continue to expand and scatter the matter and energy so thinly that mattergy will not be able to reassemble to form new galaxies. Getting something out of nothing, however, smacks of supernaturalism and defies logic, so some modern day inflationists say that there was matter in highly compressed form before the bang. Just as the bangers have had to modify their position, so have the Steady Staters who, in the modified Quasi-Steady State theory, claim that there were many smaller bangs instead of one Big Bang, but from the smaller bangs, matter and energy are recycled to make the universe safe from extinction. Steady State theorists, however, do not deny that the universe is expanding but that it is expanding at about the same pace that new matter is being created so that the density of the universe remains the same and thus appears basically the same from one era to another. Thus, the idea of creation of new matter in Steady State theory contradicts the conservation law just as the Big Bang does – unless the new matter represents a condensation of photonic radiation.

As Hegel's philosophy predicts, there has been a thesis>antithesis>synthesis with both sides making concessions to the other. However, for the time, it appears that the Big Bangers have won the day despite having two Achilles heels of major theoretical weakness: one heel based in relativity and the other in quantum mechanics which contradict each other on various levels. The following is a scenario of how Big Bangers won the war by winning key battles.
1) Steady Staters could not provide a strong counter to the evidence for the expanding and accelerating universe and when bangers provided evidence that quasars were clumped together at great distances, thus indicating that the universe was evolving rather than being in a state of relative equilibrium, the steady staters could not answer the challenge at the time. Furthermore,

when the CMBR (Cosmic Microwave Background Radiation) was discovered in the mid-60s suggesting that there was radiation left over from the Big Bang spread fairly evenly in all directions, the Steady State theory was abandoned by all except a few diehards (Mitchell 2002: pp. 257, 258).

2) Quasi-Steady State theory, the modified form of Steady State, continued to hold the idea that there were several smaller bangs but not one Big Bang. The smaller bangs would account for the aggregation of galactic groups separated by vast volumes of intergalactic space. Small bangs would produce a lumpy universe like the one we see with island galaxies, and a Big Bang would seem to produce a smooth universe with an even distribution of matter. After all, it is postulated that the heavy elements essential for life on earth came from the smaller bangs of numerous super novae. Therefore, the death of stars in supernovae leads to the birth of new stars via recycling. However, Big Bangers claim that smaller bangs could not produce the rather even distribution of the CMBR and that with smaller bangs, the CMBR would be lumpy and uneven. To refute the idea of a runaway, accelerating universe proposed by the bangers, Quasi SS theorists say that strong gravity (as predicted by Einstein) is the cause of the red shift rather than inflation. Furthermore, the spewing of hydrogen and other particles from stars and galaxies into intergalactic space could also help account for the red shift phenomenon because light from distant galaxies would be filtered to the red end of the spectrum by intergalactic hydrogen and dust. Intergalactic filtering would account for the fact that the more distant the galaxy, the faster it seems to be traveling away from us. It would appear that the greater the distance of the galaxy, the more filtering of light toward the red end of the spectrum would occur thus making the more distant galaxy appear to be traveling faster than closer galaxies. However, Big Bangers contend that filtering short waves of light does not cause the spectral lines to shift as they do with the Doppler Effect and the gravitational pull on light. The bangers also contend that the red shift due to gravity is too small to account for the entire red shift and thus the lion's share of the red shift is due to expansion of space.

Red Shift – caused by expansion or gravitation?

The red shift discovered by Hubble was perhaps the first evidence for an expanding universe. The red shift is to light what the Doppler Effect is to sound, and many refer to the shifting frequencies of light related to directional velocity as the Doppler Effect. Now there have been several explanations of the red shift other than the expanding universe. There are actually three major explanations of the red shift: 1) Tired light, 2) gravitation, and 3) expansion of the universe. Tired light, which means the filtering of light through intergalactic and interstellar dust and hydrogen, has been ruled out by some cosmologists in that it is said not to produce a shift in spectral lines – it only filters out the shorter frequencies leaving the longer frequencies in the red part of the spectrum emanating from distant galaxies to reach observers on earth. Gravitation, on the other hand, does produce a shift in the spectral lines as Einstein predicted and as Pound and Rebka demonstrated. However, orthodox cosmologists contend that the gravity of the earth produces only a small shift and is not sufficient to explain the rather large shift in frequency that is observed. Nevertheless, when the light enters the gravitational field of the Milky Way galaxy, there is another pull by the entire galaxy which lessens the frequency of light. Assuming that the gravitational pull back from the emitting galaxy causing a shift toward the red is about equal to the gravitational pull forward of the Milky Way increasing the frequency of light toward the blue, the stretching and shrinking of frequency would even out.

Of course, to admit that gravitation contributes a significant amount to the observed red shift is to weaken the inflationist view that the expansion of the space is accelerating and even surpassing the speed of light. When stating that galaxies are exceeding light speed, cosmologist have to find a patch for this violation of the universal speed limit by saying that is it space expanding faster than light – not the galaxies. According to Mitchell, a much more rational explanation than faster-than-light space expansion is that there is something besides the cosmological inflation and the Doppler Effect that is causing distant galaxies to appear to be moving away from us faster than light. That something is probably the gravitational red shift. Mitchell states that a typical neutron star, having a mass about the same as that of the sun, but only about 10 miles in diameter, would have a gravitational red shift of about 0.18, or about 90,000 times that of the sun (2002: p. 154). Mitchell goes on to show that quasars can cause significant gravitational red shift perhaps because they are associated with a Black Hole. Thus, it is not enough just to compare the gravitational pull of the earth and our galaxy on the frequency of light. One has to look at the specific source of the light and its gravity.

The effect of gravity on light frequency depends upon the direction of light in reference to the gravitating body. This point perhaps accrues to the favor of the Big Bangers, but the universe is the way it is despite the way we want it to be. Pound-Rebka found that if light is traveling toward a gravitating body, it is blue shifted as the body pulls the light toward itself. On the other hand, if the light is traveling away from a gravitating body, it is red-shifted as the body pulls back on the light escaping its grasp. Furthermore, a light beam traveling at an angle to the gravitating body will have its path curved toward the body as in gravitational lensing. Thus, it would seem that when light is leaving a star or galaxy it is being pulled opposite the direction of its motion and would manifest a red shift, but when the light is traveling toward a planet or other body, it would be blue shifted because it is being pulled toward that body. Therefore, the net shift would be determined by the size and relative gravitational strengths of the receiving and sending bodies. If the two bodies had equal gravity, it would seem that the red shift and blue shift would cancel so far as the gravitational effect is concerned. However, if one body were larger and exerted greater gravitational force than the other, then there could be a red shift or blue shift due to gravity. Additionally, the mattergy existing in intergalactic space must be considered as well. Light traveling through many lightyears of space would encounter a considerable amount of cumulative gravity; because even though the density of matter is very low in outer space, the distance is very great, and the amount of gravity would accumulate over distance and time. However, if intergalactic matter is spread evenly, then the blue-shift effect of gravity would equal the red-shift effect because there would be the same pull back as pull-forward as light traverses space. Moreover, there would be many tugs along the way from the long arm of gravity from galaxies and from stars once the light enters a particular galaxy.

Gravity's Effect on Frequency Shift

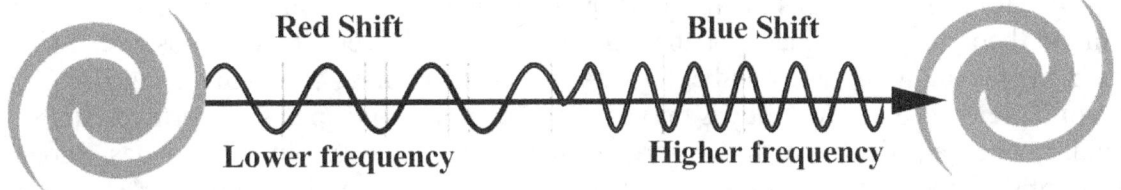

Light emitted from a galaxy is red-shifted by that galaxy as its gravity pulls back on the light. Light received by another galaxy is blue-shifted as it is pulled toward that galaxy by gravity. Intergalactive and intersteller mattergy (matter-energy) even though low in density must exert some gravitational and filtering effect on light over long distances.

Mitchell's "Recycling Universe Cosmology" (RUC) The ultimate synthesis without absurdities

Having read the various cosmological versions that attempt to explain the status of the universe, I find that William C. Mitchell's is the most rational and agrees most with my rejection of relativity and quantum theory as the mechanisms that have created the universe that we observe. Mitchell believes in the recycling of matter and energy but not in the manner of the Big Bang/Big Crunch theorists of a previous era. He believes that as one galaxy dies and spews its particles and radiation into intergalactic space, the mattergy of the old galaxy rises out of the ashes, Phoenix like, and forms a new galaxy. His idea is similar to earth ecology which involves recycling of organic matter. In this recycling process, the death of one organism provides the raw materials for another organism to be born and thrive for a while. Of course, for this recycling to occur, the universe must not be expanding so fast that the density of mattergy is so low that gravity cannot gather it back together into galaxies and stars. Hence, he believes that the universe is probably eternal and will not die a heat death due to matter being converted to radiation and scattered so widely that it is impossible to put humpty-dumpty back together again. In order to hold this view of an eternal universe that is not expanding (or not expanding very much), Mitchell has to refute the conventional Big Bang interpretation of the red shift, the Cosmic Microwave Background Radiation, the preponderance of light elements in the universe, etc. His idea of recycling of mattergy is similar to the Steady State theory; however, the Steady State theory acknowledges expansion, but basically RUC does not. Furthermore, if stars recycle their material in our galaxy, it seem reasonable that whole galaxies can recycle their materials in the process of forming new galaxies. I will attempt to summarize how Mitchell refutes each of the orthodox views of evidence for the Big Bang.

1) **Red Shift:** Mitchell believes that the red shift is caused by "tired light" and gravitation, not the expansion of space since the idea of space (nothingness) expanding is an oxymoron. The tired light is a result of the filtering of light by the pervasiveness of hydrogen and other radiation in space. Of course, standard BB orthodoxy denies that tired light and gravitation are significant factors indicating that filtration of light does not produce a shift in spectral lines.

2) **Cosmic Microwave Background Radiation**: The CMBR according to Mitchell is the result of galaxies spewing their mattergy into intergalactic space and new galaxies being born from this refuse. Bangers argue that the CMBR would not be so smooth and even and its temperature so evenly spread at 2.7K - if the radiation was coming from island galaxies. Mitchell argues that empty space has no heat – only matter has heat – so the only way that intergalactic space could be thermalized is for galaxies to eject the radiation into space at velocities necessary to escape the gravitational grasp of the galaxy. Only then could the mattergy of intergalactic space be heated up to 2.7 K. Thermalization would thus account for the smoothness of the CMBR in space. Add to this, to the fact that galaxies are older and live longer than originally thought so that there has been sufficient time for galaxies to suffuse their material into space.

3) **The distribution of elements in the universe** – the Hydrogen-Helium Ratio: In the standard Big Bang model of nucleosynthesis, the hydrogen-helium abundance helps model the expansion rate of the early universe. If the expansion had been faster, there would be more neutrons and more helium. If expansion had been slower, more of the free neutrons would have decayed before the deuterium stability point and there would be less helium (Boesgaard & Steigman 1985). Mitchell refutes this assertion by citing the many contradictions that exists in the estimates of the percentage of hydrogen. The estimates range from 20% to 25% to 36% Helium with

Hydrogen making up the lion's share, leaving about 2% for the heavier elements. Furthermore, these figures have been retrofitted to match the data, so they could hardly be called predictions. They are predictions in the same sense that Einstein "predicted" the precession in the orbit of Mercury, only after knowing the number derived from observation. Moreover, some bangers claim that hydrogen could only have come from the Big Bang, but other astronomers say that there could be other sources of hydrogen. Some galaxies have been observed to have less Helium than observed in the rest of the universe, and it is known that the fusion of Helium from Hydrogen occurs in stars such as our sun and in galactic cores, and heavier elements are said to be synthesized in super novae and spewed out into interstellar space in the explosion (Mitchell 2002: pp. 172-173).

Summary Critique of Big Bang Cosmology
My purpose here is not to say whether the Big Bang or some modified version of the Steady State Theory is true or whether the universe had a beginning and will have an end or whether it is eternal. Such issues are a matter of speculation and perhaps beyond the reach of science because, to arrive at an answer, cosmologists have violated the scientific method and resorted to metaphysics and mysticism, all the while claiming that they have not engaged in non-scientific thinking full of contradictions and absurdities. Big Bangers have seemingly reversed themselves on this issue of whether the universe is eternal or limited in time. In the early days of Big Bang theory, cosmologists claimed that there was a definite beginning of the universe and that it sprang from nothing (*ex nihilo*) out of a singularity. Now inflationists are saying that the universe didn't spring from a singularity of nothing, but that before the Big Bang, the matter of the universe was compressed into a very small space – hence there was matter, space and time before the bang contrary to the original version. Therefore, inflationists are admitting that there was indeed something before the bang and, in doing so, are implying that the universe is eternal since the bang didn't come out of nothing after all. Early bangers said that space and time along with matter and energy were created in the bang; now inflationists say that there was already space where matter was highly compressed (similar to a Black Hole), and since there was a time when something existed before the bang, time was not created by the bang. Early bangers said that the Big Bang was an explosion from a single point in spacetime and the universe expanded from that point implying that the point of explosion in space was the center of the expanded universe. Now inflationary bangers say that the Big Bang was not an explosion at all, but rather a space expansion, not from a single point, but from infinite points because space everywhere began to expand simultaneously so that there is no center to the Big Bang universe and the expansion continues at every point in space to this day. Of course, if there was no explosion of matter-energy, how would inflationist account for the extreme heat of the early universe now evidenced by the temperature of the CMBR? Then there is the conservation of mass-energy law that must be circumvented. The violation of the conservation law was a problem for early bangers who claimed that the universe came from nothing but a singularity. Now inflationists get around the problem by saying there was matter and energy before the bang. Here are some additional points of critique.
1) As discussed previously, my main critique of Big Bang Cosmology is that it is based on two flawed theories (relativity and quantum mechanics) which contradict each other on key points regarding the nature of space and time, the speed limit of light, simultaneity, causality, uncertainty, etc. While the most mystical theory of quantum mechanics (Copenhagen interpretation) is the most widely-accepted theory, there are more rational quantum theories such

as Bohmian mechanics which also has issues but not as many as the Copenhagen interpretation. Relativity, which treats space and time as physical entities that can be curved, bent, shrunk or expanded is a logical fallacy involving reification of the abstract. It is a confusion of physics with metaphysics.

2) The idea that galaxies are traveling away from us faster than light without violating the speed limit of light is a drama in the theater of the absurd. The notion that space in its expansion can violate the speed limit because space is nothing and can carry matter with it faster than light without the matter violating the speed limit is heaping the absurd upon the ridiculous.

3) **Far away galaxies are traveling away from us faster than closer galaxies**. My critique of the *farther-the-faster conundrum* is that there is something more than the Cosmological Doppler Effect due to expansion occurring. Let's imagine that there are intelligent beings in other galaxies who are observing the red shift of the Milky Way galaxy from their perspectives. One observer lives in a galaxy relatively near to us and the other lives in a galaxy far, far away. Now the far observer sees the Milky Way receding from her much faster than the close observer sees the Milky Way receding from him. If they could communicate with each other by simultaneous quantum entanglement, they might argue as to how fast the Milky Way is traveling through space since the far observer sees the recession as very fast and the closer observer sees it as much slower. They cannot both be right, so they probably would conclude that there is something besides recessional velocity or space expansion that is causing the greater red shift for the more distant observer.

4) **Special Relativity contradicts General Relativity in Big Bang cosmology**. The Big Bang based on General Relativity indicates that space is expanding carrying the galaxies faster than light. Special relativity indicates that when something is traveling near light speed, space is contracting to compensate for time slowing down in order to maintain the speed limit of light as inviolable. If something is traveling at light speed, then space and time theoretically reduce to zero – so that the object is getting nowhere fast. If an object is traveling faster than light, Special Relativity tells us that time is going backwards and therefore space would have to be going into negative numbers, that is, space would have to shrink to less than zero (if one can imagine negative space). But the absurdity of Special Relativity is that the same space can be shrinking greater for some observers than for others depending on their points of view, and the same space might not shrink at all for others at rest. Now how can the same space be expanding because of General Relativity and shrinking because of Special Relativity. To say that the same space and time can be different for different observers depending on their speed and the strength of the gravity field in which they are immersed creates tremendous cognitive dissonance to the rational mind. Such irrational theories can only be accepted on faith in mysticism. They are surely not based in science.

5) **If distant galaxies are traveling away from us faster than light, then we would never see their light.** One argument against this concept is that earthlings could see the speeding galaxy before it reached light speed. However, the counterpoint to this argument is that at the moment the astronomer looked at the light of the galaxy, the extreme red-shift indicated that the moment it was emitted, the galaxy was already moving away from us faster than light.

> **Nutshell:**
> 1) There is something wrong with a theory whose proponents claim that the high red shift indicates that galaxies are traveling away from us *faster than light*.
> 2) The fix that says that it is space that is expanding faster than light but that the matter carried

> by expanding space is not violating the speed of light is an absurdity.
> 3) The only logical resolution of this contradiction is that there is something besides the Doppler Effect due to space expansion that is causing this apparent faster-than-light recession.
> 4) If galaxies are moving away from us faster than light, then we would not see them.

Physics, the New Religion??? A Multiverse Tied Together with Strings

Greene (2011), Gribbin (2009) and others tell us that the implications of the non-scientific theories called String Theory and the Many Worlds theory indicate that there is not just one universe, but there are many universes – in a word, a multiverse. For the purpose of this analysis, we will ignore the semantic problem of having a multiverse made up of many universes. Obviously if "universe" means one, all-encompassing entity, a universe cannot be a subset of something larger. Furthermore, Gribbin tells us that there is no creator God, but there are super-intelligent beings that have created and engineered some of these fine-tuned universes such as the one we live in. Allow me to sketch this creation story in telegraphic style so as to condense it. I shall tell it verse by verse in the manner of a best-selling book. First, it is important to define some terms that are necessary for our understanding of the new religion.

~**Religion:** a set of beliefs about invisible, supernormal beings or forces that cannot be proven. These beliefs, while backed by some evidence, must be accepted as a quantum leap of faith.
~**Quantum foam**: this belief in physics goes by several aliases, such as vacuum energy, the energy of empty space, zero point energy, the energy of nothing – it is the energy inherent in so-called "empty" space.
~**Quantum fluctuation**: although the energy of the vacuum is usually spread evenly throughout space, sometimes, by chance, there is a concentration of this energy in a small volume of space. This chance fluctuation, if large enough, can result in a Big Bang and convert to matter.
~**Dark energy**, which is the same as quantum energy, is the force that is counteracting gravity in causing the universe to accelerate in its expansion – some say faster than light. It is a form of heat that causes expansion. Others say it is negative mass which produces negative gravity.
~**Cosmological Constant:** First proposed by Einstein to explain what he believed was an eternal, static universe, it is a repulsive force, that according to Einstein, balances the attractive force of gravity so that the universe does not collapse under the force of gravity and instead maintains a steady state. Einstein later said that this proposed constant was the greatest mistake of his career, but now cosmologists have seized upon it, not to argue that the universe is balanced between gravity and repulsive energy, but to explain why the universe is expanding under this repulsive force which is stronger than gravity on the intergalactic scale. Of course, that force could be none other than quantum energy in space which expands like heat and expands space when it pops into existence thus driving the universe apart.

Now, that we have linked all these aliases into one identity, on to the story, verse by verse. Gribbin, Greene and others say that eternal inflation, many worlds, dark energy and string theories all point to a multiverse, rather than a universe.

1>In the beginning, there was no beginning. The multiverse always *was* in some form because there was always quantum energy in empty space. This contradicts the original Big Bang theory which indicated that the bang started everything – space, time, matter and energy.
2>Because of a quantum fluctuation or concentration of quantum energy in empty space, a Big

Bang was created, spawning a universe. String theory indicates that the smallest particles in space are composed of even smaller strings which can be straight or form loops and vibrate at different frequencies to form particles and energy.

3>Because of dark energy (which is the same as quantum energy in empty space), the universe continues to expand because there is some heat in empty space and heat causes expansion.

4>However, despite dark energy's repulsive force, some matter, under the influence of gravity and dark matter, was able to coagulate and form galaxies with stars. But, some of the stars burned up their nuclear fuel whose heat once kept the stars stable and balanced with gravity. When the heat-generating repulsion lessened, gravity took over and compressed the remaining matter toward a singularity into a black hole - similar to the singularity that existed before the bang.

5>Smolin says these black holes are like blisters on the smooth surface of a universe and they can break off to form new universes. However, the new baby universe remains connected by a tunnel in spacetime to the mother universe, and this umbilical cord enables communication between the two. Later when the baby universe grows up, it can bubble off black holes that create new universes, and so new universes are constantly being born. Let's again ignore the semantic problem of a black hole as the most condensed form of matter in the universe being compressed to a singularity or near singularity. How can super dense matter be conceived as a hole? Ordinarily, a hole is the absence of matter. Yet we are told that a black hole really is a hole – it is a tunnel in spacetime that connects one universe to another – despite the fact that all the matter is compressed to a near singularity.

6>Now Gribbin rejects the notion, held by Garrett and others, that we live in a matrix created by a computer simulation – Gribbin believes that our universe is real. He agrees with Deutsch that there would be glitches in the simulation which we intelligent beings would be able to detect. I might add that a computer simulation involves electrons producing images with photons. Now photons are hardly like solid matter that we encounter in our empirical world. Thus to believe in a photonic universe means that the solid matter we experience is an illusion – a very Eastern, mystical concept.

7>Instead of intelligent designers that create these digital simulations, Gribbin and others say there are intelligent designers that created real universes in the material sense.

8>Now those who believe in an omnipotent God as the intelligent designer, are weak minded, but those who believe in super-intelligent aliens who create universes are highly intelligent people. So, there is no God (big G), but there are gods (little g) - sort of like demi-gods. In the sense that Von Daniken believes that super-intelligent aliens built the pyramids and other megalithic structures which humans did not have the intelligence to build, so super intelligent beings, who evolved according to Darwinian evolution in other universes, created our universe and others. This is a very Buddhist idea – in that there is no one creator God, but there are Bodhisattvas and divas who take on the status of demi-gods.

9>Now, these super beings made universes by creating black holes which broke off to form new universes, but as in Darwin's theory, each new universe had some mutations so that the constants and natural laws of the baby universe were slightly different from the mother universe. Some of these mutations were conducive to life (anthropic principle) and some were not. At any rate, after the baby universe was created, it had to develop on its own without help from its mother. That means that Darwinian evolution, based on random mutations, would be the way life evolved in the new universe – there would be no designer present who would direct this specific process – life would develop naturally on its own. This idea is rather like "Deism" in which God, like a watchmaker, created the universe and its laws and left it alone to develop on its own without further intervention.

10>However, as the super-intelligent beings became more intelligent, they could engineer the constants and laws and fine tune them to make it possible for more and more intelligent beings to evolve and better universes to form. The super-dupers would produce *designer universes*, if you will – similar to Hitler's attempt to design a super race.

11>Now the way super beings designed a more intelligent universe was to manipulate the tiny strings that make up all matter. You see, the shapes of the 6 or 7 extra dimensions in String Theory determines how the strings can curl up into loops which in turn determines how the strings can vibrate and form particles such as quarks, protons, neutrons, electrons, etc. As an analogy, think of how the *shape* of a trumpet, trombone, French horn or a bass tuba can determine the pitch of sound when air vibrates as it is blown through these instruments. So, in another universe, these curled up spaces might make a proton with two quarks instead of three – and that might generate a different form of life. There are 10^{500} ways that the extra dimensions can shape strings, so that means there are at least 10^{500} types of universes that can be formed with different constants and different types of particles that comprise macro matter. And, each type of universes can contain many universes of that type, so the number of universes in the multiverse could be infinite (but then infinity is not a number).

12>Christians and other theists say that the universe has been *designed* which implies a designer. Gribbin and company agree and say, *yes, the universe we live in has been designed* with fine-tuned constants and laws by super-duper aliens, so that it was inevitable that life would develop, but life still developed *by chance* in Darwinian fashion. In other words, life was predestined to evolve by chance.

13>This imitation of deism enables Gribbin to say that there are intelligent designers of our universe (which means it didn't come about by chance), but cling to the Darwinian idea that life in our universe did indeed arise by chance. Yet, the designers created our universe in such a way as to make life inevitable (not based on chance). In other words, life was predestined, but it still arose by chance. How does one predetermine something specific to come into being by chance? Sounds like the logic or illogic of hedging to me.

14>Yet, it seems that if the super-braniacs of the multiverse could create whole designer universes, they would surely know something about biology and could create a *designer species* that would be stronger and more intelligent. Even we less-intelligent humans know how to direct evolution in non-chance ways to breed animals so as to get super-fast race horses or super-strong dogs. Humans also know how to recombine DNA of different species to create a chimera (such as splicing frog genes into tomatoes to give them a longer shelf life), how to clone sheep and other animals, how to create mutations with radiation, how to change some genes, etc. Are we lowly humans smarter in biology than the creators of our universe? Or is it that Gribbin does not want to acknowledge intelligent design in biology by aliens because that would score a point for the Christians and other theists who believe that evolution did not produce intelligent humans by chance alone. Scientists have their biases too. As a friend of mine said, materialist scientists need the multiverse in order to support atheistic-materialism. If there are enough universes with varying constants and laws, then the multiverse casino would inevitably hit the jackpot and come up with the combination of factors necessary for life. But, then, Gribbin and others still think that there needs to be intelligent designers to load the dice to make life likely, if not inevitable – as in the anthropic principle.

So is physics the new religion? Smolin, Penrose and Woit tell us that String Theory is not a scientific theory because it produces no predictions or hypotheses that can be tested empirically – empiricism being the hallmark of science. Similarly, the Multiverse theory, largely based on String Theory, cannot be tested empirically and is unscientific. Since String Theory, the Multiverse and the Many Worlds Theory cannot be tested scientifically, they all require a quantum leap of faith and mathematical imagination to believe. Is this leap of faith different from theists who believe the evidence of design in the universe points to a super designer? Yours truly doesn't think there is any difference. So, again, I pose the question: Is physics, in its modern form, the new religion? Admittedly, some physicists do not believe that String Theory or the Multiverse Theory represent real science. Perhaps it could be described as mysticism rather than religion.

More critique of String Theory and the Multiverse

1>The language of space and time is the same language Einstein used in relativity, i.e., space is a material thing like matter and energy that can determine the properties and behavior of matter and energy. This language is seen in describing the additional dimensions of space needed for String Theory to work. Because these spaces cannot be observed, the hedge that is used is that they are compacted and therefore too small to be seen. Furthermore, these spaces must be compacted and given different shapes in order to produce the different frequencies of looped strings trapped in these dimensions. *Now the idea that space has any shape at all is a fallacy.* Space defined as emptiness or void has no shape except that given to it by matter that surrounds it. For example, if a sealed cube is evacuated in intergalactic space, the empty space in the box could be described as a cube, not because of the shape of space, but the shape of matter that contains it.

2>Likewise the manifold (many-folded), curled up spatial dimensions are said to have holes that the strings can attach to. **Space *is* a hole,** so how can a hole have a hole in it? A hole can only exist in a material object – a hole is a space in an object where there is no discernible matter. If I could knock a hole in a wall in intergalactic space, the hole is where there is only space (no matter, except possibly the quantum foam which involves virtual particles).

3>Superstring theory is a hedge for String theory. When String theory hit a wall with many contradictions, it needed a hedge, so String theorists could dodge the inconsistencies that had developed. There were five different String theories when there was supposed to be only one. So, they turned to mathematics and Edward Witten's M-Theory (perhaps the Mother of all theories) and found Supersymmetry (SUSY) which gave them Superstring theory. However, SUSY did not solve their problems. You see, String theory was supposed to be the Big TOE (theory of everything) that was purported to unite all the forces of nature including the difficult one, gravity, into one single, elegant equation. SUSY was also supposed to unite all the particles in the Standard Model, and SUSY told us that for every mass particle (fermion), there was a symmetric massless particle (boson) and vice versa. Thus, every fermion has a supersymmetric particle called a sparticle. A symmetric particle is one that looks just like its super partner in the mirror, so it is reversed and differs in one property. Instead of one beautiful equation, however, SUSY has given us 10^{500} equations. Where has all the unity gone, long time passing? Not a problem, says Susskind and other players in his String band, because that huge number (greater than the number of particles in the known universe) describes all the types of universes that String music can give us. And, our universe can be described by *only one* of those equations, giving us the holy grail of grand unification for our universe. See, we can't unify the multiverse with strings, but we can unify our little universe with strings, so untied strings are now tied and united.

The other $10^{499.999...}$ equations describe the other types of universes made by vibrating strings in curly, cramped dimensions. The only problem is that we have no way of knowing which one of the 10^{500} equations describes our local universe. However, we do know that given an infinity of universes, there are many twin universes (really, just like ours) where you and I have as many doppelgangers as there are universes that were created with our unique equation. Too bad, we can never meet our many twins, or triplets or quadruplets, or quintuplets *ad infinitum*.

4>M-Theory (Mother theory) also provided a unification for all five of the separate String theories. Mother says that these five strings are just manifestations of the one string. You see, all the strings are really attached at right angles to a single 2-dimensional brane (from mem<u>brane</u>). It is the 2-D brane that is vibrating causing the strings attached to it to vibrate like a sheet flapping in the wind. Then branes can be stacked one over the other, and these branes can be attached to each other by the strings tied at both ends. Universes then are like little holes in the branes – like the holes in cheese or perhaps more like a mold spot between two slices of bread. One of the problems with the 2-D brane theory, which doesn't take a lot of brains to figure out, is that there is no flat, two-dimensional matter – all observed matter is 3-D and has thickness as well as length and width. How could a string attach to anything that doesn't have any thickness to hold it? Of course, such reasoning is not allowed in the realm of the small (quantum) or the large (general relativity) where things are totally different from the middle, mundane 3-D world we live in.

5>There is the assumption that the constants and laws of nature can vary from one universe to another. While that may be true, empirically, we know of only one set of constants and laws – the ones we observe in this universe. Perhaps to make a proton, three quarks must be assembled in a certain way, and that would be the same in other universes (if there are other universes). Another assumption is that there can be only one form of life and that is the type that resides in our little corner of the universe. So, if other forms of life are possible (even in our local universe) using a different chemistry, then it does not appear that the constants were designed just with us in mind. The same constants could have created a non-carbonated chemistry that spawned other forms of life. Of course, that would weaken the case for a designer universe, and, instead, life developed from what was there, rather than the universe being a "put up job" so that life of our unique type would inevitably develop.

Failure to Unite forces leads to many epicycle-type hedges in String Theory Hedge Fund

1>The original intent of String Theory was to unite all the forces of nature into *one* elegant equation. Instead, String Theory produced 5 different equations.

2>The hedge for this failure was M-theory developed by Edward Witten who claimed that the five different equations could be united into one grand theory using Supersymmetry and Calibi-Yau compacted spaces. And, why did the 6 or 7 extra dimensions have to be compacted into very small spaces? Because another hedge was needed to explain why we can't see into these extra dimensions.

3>However M-theory with its Calibi-Yau compacted spaces of various shapes and string holes produced 10^{500} equations, not one unique equation. The situation got worse, not better – the equations went from 5 to 10^{500}.

4>The hedge for this failure was that Susskind and others contended that these 10^{500} equations describe a multiverse and that each universe (including ours) within the multiverse is described by one of these equations. The problem is - it cannot be determined which one of these constants describes our universe.

5>With 10^{500} equations describing the cosmological constants of myriad universes, almost any theory about the multiverse can be confirmed. In other words, the theory is insulated from falsification.
6>Hence this theory of everything (TOE) is a theory of nothing (TON) because almost anything goes.

Here's what Paul Davies has to say about having so many possible universes (with so much variety in constants) that almost anything one could make up about a universe could be true somewhere. Any such hypothesis would be virtually unfalsifiable.

It is not too much of an exaggeration to say that you could dream up a universe, choosing whatever sort of low-energy physics takes your fancy (within reason), and there will be a universe somewhere matching that description among the unimaginably vast smorgasbord of possibilities (Gribbin 2009: p. 170).

Thus superstring theory is a huge hedge fund that fails to unite all the different String theories, the particles of the Standard Model with their super partners, quantum theory with general relativity, and gravity with the other three forces – because it produces no testable predictions and is unscientific. Actually, if gravity is not a force, but merely grooves in spacetime that compel smaller bodies to orbit larger bodies, as Einstein claimed, non-gravity grooves would not need a particle to carry its force. As you can see, there is no longer a boundary between science and science fiction, physics and metaphysics. See the discussion of String Theory in my Quantum Theory book for more information on Strings.

COSMIC SENSE, COMMON SENSE...

SUMMARY

Cosmology borrows metaphysical and mystical baggage from quantum theory and relativity to explain large-scale, universal phenomena. The singularity in spacetime (a dimensionless point) is borrowed from General Relativity and is obviously a metaphysical construct since matter-energy cannot logically exist in non-space and non-time. The mystical idea of getting something from nothing is borrowed from quantum physicists who inform us that virtual particles can pop into existence out of the vacuum (which has zero-point energy) and can become real particles of matter under certain circumstances. This quantum foam, as it is called, gives a rationale for how the Big Bang could have banged into existence from nothing.

The idea of space expansion comes from relativity which indicates that space is a flexible physical medium. However, the notion in cosmology, which contradicts relativity, is that space is expanding faster than light, moving galactic islands from each other at greater and greater distances. The patch for the faster-than-light space expansion contradiction is that matter is not moving faster than light because the matter in galaxies is at rest in space and is just being carried along at superluminal speeds. As Michio Kaku says: "Space is *nothing* and *nothing* can go faster than light." Of course, Kaku is missing the key ingredient in relativity, namely, that space is *something* that can be bent curved, warped, twisted, folded, stretched and contorted in many different ways.

There is much contradiction in whether anything can escape a black hole. Initially, physicists told us that nothing, not even light or electromagnetic radiation, can escape a black hole. Then Hawking found a way that mass can be destroyed in a black hole, thus leading to its dissolution. He contends that anti-matter particles hatched with their matter twins in the quantum foam can be pulled into black holes and annihilate an equal amount of matter. However, we have found that matter is not destroyed in these matter-antimatter collisions – but is converted to radiation, which can be converted back to matter. Therefore, the radiation created from matter-antimatter collisions cannot escape from the hole if the original contention is true – that not even light can escape its clutches. Thus, there is still no explanation as to how radiation escapes and dissolves the black hole – if indeed it does.

References

Abbott, B. (2007). "Microwave (WMAP) All-Sky Survey". Hayden Planetarium.

Aharonov, Yakir and Rohrlich, Daniel (2005) *Quantum Paradoxes: Quantum Theory for the Perplexed.* Wyley VCH

Aharonov, Yakir (2002) *Uncertainty.* Discovery Science Video.

Albert Einstein Site Online. Retrieved from http://www.alberteinsteinsite.com/quotes/einsteinquotes.html.

Alok, Jha, (August 6, 2013). "*One year on from the Higgs boson find, has physics hit the buffers?*". The Guardian: London.

Ananthaswamy, Anil (2017) *A classic quantum test could reveal the limits of the human mind.* New Scientist. https://www.newscientist.com/article/2131874-a-classic-quantum-test-could-reveal-the-limits-of-the-human-mind

Ashby, Neil (2002) "Relativity and the Global Positioning System." *Physics Today*, May 2002, p. 41.

Barish, Barry C. and Weiss, Rainer (October 1999). "LIGO and the Detection of Gravitational Waves". Physics Today. 52 (10).

BBC Documentary (2008) *Hawking Radiation*. Interview with Professor Bernard Carr, Queen Mary University of London. https://www.youtube.com/watch?v=S6srN4idq1E

Beckmann, Petr (1987) *Einstein Plus Two.* Boulder, CO: Golem Press.

Behe, Michael J. (1996). *Darwin's Black Box: The Biochemical Challenge to Evolution.* New York: Free Press.

Berman, Bob (Dec. 2017) "Intelligent Design" in *Astronomy* Magazine.

Bethel, Tom (2009) *Questioning Einstein: Is Relativity Necessary?* Pueblo, CO: Vales Lake Publishing.

Biever, C. (6 July 2012). "It's a boson! But we need to know if it's the Higgs". New Scientist. Retrieved 2013-01-09.

Bishop, Owen (1984) *Yardsticks of the Universe.* New York: Peter Bedrick Books.

Boesgaard, A. M. and Steigman, G. (1985). "Big Bang Nucleasynthesis: Theories and Observations", *Ann. Rev. Astron. and Astrophys.* 23, 319.

Boswell, J. (1823). The Life of Samuel Johnson, vol. 1. London: J. Richardson & Co.

Bohm, David (1980) *Wholeness and the Implicate Order*. New York: Routledge & Kegan Paul.

Brainy Quotes: http://www.brainyquote.com

Brandenburg, John (2011) Beyond Einstein's Unified Field. Kempton, IL: Adventures Unlimited Press.

Brown, James Cooke (1975) *Loglan 1: A Logical Language*, Loglan Institute.

Bryce, Emma (2016) *Will Wind Turbines Ever be Safe for Birds*. http://www.audubon.org/news/will-wind-turbines-ever-be-safe-birds

Bunge, Mario (2001). *Philosophy in Crisis: The Need for Reconstruction*. Amherst, New York: Prometheus Books.

Capra, Fritjof (1999) *The Tao of Physics*. Boston: Shambhala Publications.

Carrey, Jim (1995) *Ace Ventura: When Nature Calls*. Movie. Morgan Creek Productions.

Carroll, Sean (May 20, 2013) *Arrow of Time - Sixty Symbols*. https://www.youtube.com/watch?v=9VFGuupXwng#t=235.460771

Carroll, Sean (Jan 22, 2013) *Quantum Mechanics (an embarrassment) - Sixty Symbols*. https://www.youtube.com/watch?v=ZacggH9wB7Y

Carroll, Sean M. (2006). *C-SPAN broadcast of Cosmology at Yearly Kos Science Panel, Part 1*.

Carroll, Sean M. (2004). *Spacetime and Geometry*. Addison Wesley.

Case, Thomas. (2013) *A Short Introduction to Metaphysics*. (Kindle Locations 49-50). Didactic Press. Kindle Edition.

Caughill, Patrick (July 28, 2017) *A New Breakthrough in Quantum Computing is Set to Transform Our World*. Futurism website: https://futurism.com/a-new-breakthrough-in-quantum-computing-is-set-to-transform-our-world/

Cherenkov, Pavel A. (1934). "Visible emission of clean liquids by action of γ radiation". Doklady Akademii Nauk SSSR 2: 451. Reprinted in Selected Papers of Soviet Physicists,Usp. Fiz. Nauk 93 (1967) 385. V sbornike: Pavel Alekseyevich Čerenkov: Chelovek i Otkrytie pod redaktsiej A. N. Gorbunova i E. P. Čerenkovoj, M.,"Nauka, 1999, s. 149-153.

CERN Document Server, http://home.cern/topics/antimatter

Cervantes-Cota, J.L.; Galindo-Uribarri, S.; Smoot, G.F. (2016). "A Brief History of Gravitational

Waves". *Universe* 2 (3): 22. doi:10.3390/universe2030022

Chomsky, Noam (1957) *Syntactic Structures.* The Hague: Mouton.

Clegg, Brian (2012) *Gravity: How the Weakest Force in the Universe Shaped Our Lives.* St. Martin's Press. Kindle Edition. (pp. 149-150, Kindle location 1787).

Clegg, Brian (2014). *30-Second Quantum Theory.* New York: Metro Books.

Cox, Brian (2011) *The Quantum Universe: (And Why Anything That Can Happen, Does)* Da Capo Press. Kindle Edition.

Cox, Brian and Forshaw, Jeff (2010) *Why Does $E=MC^2$.* Da Capo Press, Kindle Edition.

Davies, Paul (2003) *How to Build a Time Machine.* Penguin Publishing Group. Kindle Edition.

Davies, Paul and Gregersen, Niels Henrik (2010) *Information and the Nature of Reality: Physics to Metaphysics.* New York: Cambridge University Press.

Davies, Paul (2006) Interview in: "The Anthropic Principle" Video by IBBC Worldwide Ltd.

Del Rosso, A. (19 November 2012). "Higgs: The beginning of the exploration". CERN Bulletin. Retrieved 2013-01-09.

Deutsch, David *(1998) The Fabric of Reality: The Science of Parallel Universes. Amazon Kindle Book.*

Deutsch, Sid (2005) *Einstein's Greatest Mistake: The Abandonment of the Ether.* New York: iUniverse, Inc.

Dirac, P.A.M (May 1963) "The Evolution of the Physicist's Picture of Nature," in Scientific American, May 1963, p. 53.

Discovery Science Video (viewed 2002) *Uncertainty.*

Dyson, F.W.; Eddington, A.S.; Davidson, C.R. (1920). *"A Determination of the Deflection of Light by the Sun's Gravitational Field, from Observations Made at the Solar eclipse of May 29, 1919".* Phil. Trans. Roy. Soc. A 220 (571-581): 291–333.

Eagle, Bob, aka Dr. Physics (April 9, 2012) *Bell's Inequality.*
https://www.youtube.com/watch?v=qd-tKr0LJTM

Einstein, Albert (1916) *Memorial Notice for Ernst Mach,* Physikalische Zeitschrift 17: 101-02.).

Faye, Jan (Fall 2014 Edition) "Copenhagen Interpretation of Quantum Mechanics", *The Stanford Encyclopedia of Philosophy* Edward N. Zalta (ed.).

https://plato.stanford.edu/archives/fall2014/entries/qm-copenhagen

Feynman, Richard P (1990) *QED, The Strange Theory of Light and Matter*, Penguin, p. 128

Feynman, Richard (1965) *The Character of Physical Law.*

Folger, Tim (June, 2002) *Does the Universe Exist if We're Not Looking?* Discover Magazine.

Frank, Adam (2016). *Three Atoms per Cubic Meter.* In NPR Blogger 13.7: Cosmos and Culture. Retrieved from http://www.npr.org/2016/08/09/489361654/short-answers-to-big-questions-exploring-atoms-in-space.

Gardner, Martin (1997) Relativity, Simply Explained. Dover Publishing.

Garret, Ron (2011) *The Quantum Conspiracy: What Popularizers of QM Don't Want You to Know.* Google Tech Talk. https://www.youtube.com/watch?v=dEaecUuEqfc.

Gawiser, E.; Silk, J. (2000). "The cosmic microwave background radiation". *Physics Reports* 333–334: 245–267.

Gibbs, Phillip (1996). "*Can Special Relativity Handle Accelerations?*" The Original Usenet Physics FAQ. Retrieved 2014-07-23.
http://math.ucr.edu/home/baez/physics/Relativity/SR/acceleration.html

Gingerich, Owen (1992) *The Great Copernicus Chase.* Sky Publishing Corp.http://astro.wsu.edu/worthey/astro/html/im-lab/stonehenge/stonehenge.html

Goldsmidt, Walter (1970) *Whatever Happened to Human Nature.* Lecture at Wake Forest University.

Good Reads: http://www.goodreads.com/quotes

Greene, Brian (1999) *The Elegant Universe.* New York: W.W. Norton and Company.

Greene, Brian (1999) *The Elegant Universe: Superstrings, Hidden Dimensions, and the Quest for the Ultimate Theory.* W. W. Norton & Company. Kindle Edition.

Greene, Brian (2016) *M-Theory, String Theory and the Elegant Universe.* Discovery Channel - String theory rare documentary, National geographic retrieved from Youtube: .
https://www.youtube.com/watch?v=qtaAM84Kt2I.

Greene, Brian (2004) *The Fabric of the Cosmos.* New York: Random House.

Greene, Brian (2011). *The Fabric of the Cosmos:* Video by NOVA and PBS.

Greene, Brian (2014). *The Fabric of the Cosmos: What is Space?* Youtube: Rising Life Media.

Gribbin, John R. (1984) *In Search of Schrodinger's Cat.* Bantam Books.

Gribbin, John R. (2009) *In Search of the Multiverse.* John Wiley and Sons.

Gules and Sable: *Escutcheons of Science*

Gunion, John F.; Haber, Howard; Kane, Gordon; Dawson, Sally (2008) *The Higgs Hunter's Guide.* Westview Press.

Hadhazy, Adam (2016, December) *Nothing Really Matters.* Discover Magazine.

Hameroff, Stuart (2008). "That's life! The geometry of π electron resonance clouds". In Abbott, D; Davies, P; Pati, A. Quantum aspects of life (PDF). World Scientific. pp. 403–434.

Hatch, Ronald (1995) *"Relativity and GPS,"* Part I, Galilean Electrodynamics, 6, 3, pp. 51-57, and Part II, Ibid. 6, 4, pp. 73-78)

Hatch, Ronald R (1995) *Relativity and GPS*, Part II, Galilean Electrodynamics 6, 4, p. 73-78/ http://aetherforce.com/the-suppression-of-inconvenient-facts-in-physics-by-rochus-boerner/

Hatch, Ronald (2004) *"Those Scandalous Clocks."* GPS Solutions: 67-73, p. 72.

Hayden, Howard and Whitney, Cynthia (1990) *"If Sagnac and Michelson-Gale, Why Not Michelson-Morley?"* Galilean Electrodynamics, Vol. 1: Nov/Dec.

Hawking, Stephen (2014) *The Beginning of Time.* Lecture: http://www.hawking.org.uk/the-beginning-of-time.html

Henry, Richard Conn (7 July 2005).*The Mental Universe.* Nature 436, 29 | doi:10.1038/436029a.

Herbert, Nick (1985). Notes from Quantum Reality.
http://www.basicincome.com/bp/quantumreality.htm

Hilgevoord, Jan, ed. (1995) *Physics and Our View of the World.*

Hille, Karl (March 23, 2017) *Gravitational Wave Kicks Monster Black Hole Out of Galactic Core.* https://www.nasa.gov/feature/goddard/2017/gravitational-wave-kicks-monster-black-hole-out-of-galactic-core

Hill, Paul R. (1995) Unconventional Flying Objects: A Former NASA Scientist Explains How UFOs Really Work (Kindle Locations 1261-1265). Hampton Roads Publishing. Kindle Edition.

Holzner, Steven (2004) *Physics II for Dummies*. Hoboken, NJ: Wiley Publishing, Inc.

Holmbberg, Eric (2006) *The Anthropic Principle.* Video produced by BBC Worldwide Ltd.

Hossenfelder, Sabine (2017) *No, physicists have not created "negative mass.* Backreactions blog. http://backreaction.blogspot.com/2017/04/no-physicists-have-not-created-negative.html

Hughes, R.I.G. *The Structure and Interpretation of Quantum Mechanics.* Harvard University Press.

IANDS (International Association for Near Death Studies) (2016) Report of AWARE study. https://iands.org/news/news/front-page-news/1060-aware-study-initial-results-are-published.html

Iowa State Dept. of Physics and Astronomy (2001) *Polaris Project.* http://www.polaris.iastate.edu/EveningStar/Unit2/unit2_sub1.htm.

IPN Progress Report 42-159 2004

Iverson K.E. (1980) "*Notation as A Tool of Thought*", Communications of the ACM, 23: 444–465.

Ives, H. E.; Stilwell, G. R. (1938). "An experimental study of the rate of a moving atomic clock". Journal of the Optical Society of America. 28 (7): 215.

Jaffe, R. (2005). "Casimir effect and the quantum vacuum". Physical Review D. 72 (2): 021301.

Jagerman, Louis (2001) *The Mathematics of Relativity for the Rest of Us*. Victoria, B.C.: Trafford Publishing.

Jansen, K.L.R. (1999) Ketamine (K) and Quantum Psychiatry. Asylum 11 (3) 19-21.

Jorlunde Film Denmark (1985) *Quantum Entanglement Documentary - Atomic Physics and Reality*. Published on youtube in 2014 by Muon Ray.
https://www.youtube.com/watch?v=BFvJOZ51tmc&list=PLtnb8DfCuFNx8x7Jga7K7Wni-eccc3r42

Kaiser, David (2014, Nov. 14). *Is Quantum Entanglement Real?* New York Times Sunday

Review: SR10.

Kaku, Michio (2002) Statement on Space as Nothing. Discovery Science Video and (2011) *How the Universe Works*. Video by Discovery Communications produced by Pioneer Productions.

Kaku, Michio (2011) *Michio Kaku Explains String Theory*. Youtube video: https://www.youtube.com/watch?v=kYAdwS5MFjQ

Kaku, Michio (2013) *Is God a Mathematician?* Youtube Video: Big Think Channel https://www.youtube.com/watch?v=jremlZvNDuk.

Kaku, Michio (2015) *Can universes form from "nothing"?* Youtube video: https://www.youtube.com/watch?v=JlcHMI0cC00

Kelly, Alphonsus G. PhD (2005) *Challenging Modern Physics*. Boca Raton, FL: Brown Walker Press.

Kenyon, Dean (2002) Interview in "Unlocking the Mystery of Life." DVD by Illustra Media

Khamehchi, M. A. et al (2017). *Negative-Mass Hydrodynamics in a Spin-Orbit–coupled Bose-Einstein Condensate*, Physical Review Letters DOI: 10.1103/PhysRevLett.118.155301

Kim, Y.H. & Shih, Y. (1999). "Experimental realization of Popper's experiment: violation of the uncertainty principle?" *Foundations of Physics* 29 (12): 1849–1861.

Known Universe (2009) *The Fastest*. Season One: Episode 3. National Geographic Production.

Kolata, Gina (1987) *The Sad Legacy of the Dalkon Shield*. The New York Times: December 6, 1987.

Krasnitz, Michael (2002). *Correlation functions in supersymmetric gauge theories from supergravity fluctuations hHKtions* (PDF). Princeton University Department of Physics: p. 91.

Krauss, Lawrence M. (2012). *A Universe from Nothing: Why There Is Something Rather Than Nothing*. New York: Free Press.

Kuhn, Thomas S. (2012) *The Structure of Scientific Revolutions*. Chicago, IL: University of Chicago Press.

Laureyssens, Dirk (2009).*The Gravitational ETHER of Einstein*. Retrieved from: http://www.mu6.com/einstein.html).

Lee, Penny (1996), "*The Logic and Development of the Linguistic Relativity Principle*", the Whorf Theory Complex: A Critical Reconstruction. John Benjamin's Publishing, p. 84.

Lincoln, Donald (March 13, 2018) *Twin Paradox: the real explanation.* Fermi Lab Youtube site: https://www.youtube.com/watch?v=GgvajuvSpF4

Lincoln, Donald (May 21, 2013) *What is Supersymmetry?* Fermi Lab Youtube site: https://www.youtube.com/watch?v=0CeLRrBAI60

Leplin, Jarrett (1984), *Scientific Realism*, University of California Press.

Lévi-Strauss, C. (1967). *Structural Anthropology*. Translated by Claire Jacobson and Brooke Grundfest Schoepf. New York: Doubleday Anchor Books.

Long, Jeffrey (2010) *Evidence of the Afterlife: The Science of Near-Death Experiences.* New York: HarperCollins.

Maglab (2014) https://nationalmaglab.org/education/magnet-academy/watch-play/interactive/electromagnetic-deflection-in-a-cathode-ray-tube-i

Mandelbaum, Ryan F. (4/21/2017) *No, Scientists Didn't Just Create Negative Mass or Defy the Laws of Physics*. http://gizmodo.com/no-scientists-didnt-just-create-negative-mass-or-defy-1794525465

Markey, Sean (October 8, 2003) *Universe is Finite, "Soccer Ball"-Shaped.* National Geographic News: http://news.nationalgeographic.com/news/2003/10/1008_031008_finiteuniverse.html

Martineau, Harriet (1853) *From the Positive Philosophy of Auguste Comte* (translated London) Vol. I.

Masten, Luke (2006) "Fred Hoyle." In: *The Physics of the Universe.* http://www.physicsoftheuniverse.com/scientists_hoyle.html

Maybury-Lewis, David (1992) *Millennium: Tribal Wisdom and the Modern World*. The Global Television Network: video and book.

Mbarek, Saoussen, Paranjape, M. B. (2014) *Negative mass bubbles in de Sitter space-time*. Phys. Rev. D 90, 101502(R).

Mike, John (2011) *The Anatomy of a Flying Saucer* (Kindle Location 1907). . Kindle Edition.

Milonni, Peter W. (1994) *The Quantum Vacuum: An Introduction to Quantum Electrodynamics* Academic Press.

Minick, Scot (2002) "Unlocking the Mystery of Life". Video by Illustra Media.

Minutephysics (Sep 13, 2017) *Bell's Theorem: The Quantum Venn Diagram Paradox.* https://www.youtube.com/watch?v=zcqZHYo7ONs&list=PLZ7qfdXfdJe6_0ezTgUZD2d9MBRnFAW8D

Mitchell, William C. (2002) *Bye, Bye Big Bang, Hello Reality.* Carson .City, Nevada: Cosmic Sense Books.

Moon, P. and Spencer, D.E. (1956) *"On the Establishment of Universal Time"*, Phil. Sci., Vol. 23, No. 3 (Jul., 1956), pp. 216-229.

Musser, George (2003) *According to the big bang theory, all the matter in the universe erupted from a singularity. Why didn't all this matter--cheek by jowl as it was--immediately collapse into a black hole?* Scientific American: scientificamerican.com/article/according-to-the-big-bang/

Munday, J.N.; Capasso, F.; Parsegian, V.A. (2009). "Measured long-range repulsive Casimir-Lifshitz forces". Nature. 457 (7226): 170–3.

Naeser, C. W. (1979). *Fission-Track Dating and Geologic Annealing of Fission Tracks.* In: Jäger, E. and J. C. Hunziker, Lectures in Isotope Geology, Springer-Verlag.

Nave C. R. (2016) *"Electroweak Unification".* Hyperphysics (Georgia State University). http://hyperphysics.phy-astr.gsu.edu/hbase/Forces/unify.html

Navipedia: Troposphere Monitoring. www.navipedia.net

"NAVSTAR GPS User Equipment Introduction" (PDF). US Coast guard navigation center. US Coast Guard. September 1996

Nawrot, W. (1998) "*Some remarks on Around-the-World Atomic Clocks Experiment*", Submitted to International Journal of Theoretical Physics.

Ornstein, Robert (1977) *The Psychology of Consciousness.* New York: Harcourt, Brace, Javanovich.

Ornstein, Robert (1997) *The Right Mind.* New York: Harcourt, Brace, & Company.

Orzel, Chad (2018) *The Real Reasons Quantum Entanglement Doesn't Allow Faster-Than-Light Communication.* https://www.forbes.com/sites/chadorzel/2016/05/04/the-real-reasons-quantum-entanglement-doesnt-allow-faster-than-light-communication/#699818543a1e

Pandian, Jagadheep D. (June 27, 2015) "Why is the Universe flat and not spherical?" (Advanced)

Ask an Astronomer: http://curious.astro.cornell.edu/about-us/103-the-universe/cosmology-and-the-big-bang/geometry-of-space-time/600-why-is-the-universe-flat-and-not-spherical-advanced

Parapsychological Association (February 11, 2011) *PK on random number generator.* http://www.parapsych.org/articles/36/66/1_pk_on_random_number_generators.aspx

Paul (2016) *Michelson-Morley Experiment Explained.* Youtube video: https://www.youtube.com/watch?v=F2m_VZJM0Zc

Penrose, Roger (1995). *Shadows of the Mind: A Search for the Missing Science of Consciousness.* Oxford: Oxford Univ. Press.

Penrose, Roger (1999). *The Emperor's New Mind: Concerning Computers, Minds, and the Laws of Physics.* Oxford: Oxford Univ. Press.

Petkov, Vesselin (14 December 2003) *Propagation of light in non-inertial reference frames.* Science College, Concordia University. Retrieved from: https://arxiv.org/pdf/gr-qc/9909081.pdf

Pew, Glen (2010). Supersonic Flight, Sonic Booms. AVweb Video: Retrieved from Youtube: https://www.youtube.com/watch?v=gWGLAAYdbbc.

Physics Forums (Oct.19, 2012) https://www.physicsforums.com/threads/do-gravity-waves-imply-repulsive-force-component.645331/

Pike, O, J. et al. (2014) 'A photon–photon collider in a vacuum hohlraum'. *Nature Photonics*, 18 May 2014.

Pogge, Richard (2015) *Real-World Relativity: The GPS Navigation System.* http://www.astronomy.ohio-state.edu/~pogge/Ast162/Unit5/gps.html

Popper, Karl (1962). *Conjectures and Refutations.* New York: Harper Torchbooks.

Popper, K. R. (1967), "Quantum Mechanics Without 'the Observer'", in Mario Bunge (ed.), Quantum Theory and Reality, New York: Springer, pp. 1–12.

Popper, Karl (1982).*Quantum Theory and the Schism in Physics.*

Popper, Karl (1985). "Realism in quantum mechanics and a new version of the EPR experiment" In Tarozzi, G.; van der Merwe, A. Open Questions in Quantum Physics. Dordrecht: Reidel. pp. 3–25.

Powell, Kevin (2016, July/August) *Entanglement.* Discover

Quigg, Chris (2008, January 17). "Sidebar: Solving the Higgs Puzzle". Scientific American.

Reucroft, Stephen and Swain, John (2016) *What is the CASIMIR EFFECT?* Scientific American website, retrieved 12-30-2016, https://www.scientificamerican.com/article/what-is-the-casimir-effec/#checkout

Ricker, Harry H. III *What Happened to Dingle?* http://www.mrelativity.net/Papers/18/Ricker.htm

Sears, Young & Zemansky (1999) *University Physics.* Addison-Wesley. Pp.18–38.

Shankland, R.S. et al. (1955) Rev. Mod. Phys. 27 no. 2, pp. 167–178

Sheldrake, Rupert (2009) *Morphic Resonance.* Rochester: Rock Street Press.

Shifman, Mikhail (October 31, 2012) *Reflections and Impressionistic Portrait at the Conference Frontiers: Beyond the Standard Model*, FTPI (pdf).

Shockey, Peter (2013) *George Rodonaia's – NDE – A Scientist's Afterlife.* Documentary: http://www.lifeafterlife.tv/

Smolin, Lee (2006) *The Trouble with Physics: The Rise of String Theory, The Fall of Science, And What Comes Next.* Harcourt Brace Publishers

Sofaer, A., Zinser, V. & Sinclair, R. M. (1979 a). *A Unique Solar Marking Construct.* Science, 206, 283-291.

Sokal, Alan D. (5 June 1996). "A Physicist Experiments with Cultural Studies". *Lingua Franca.*

Sorensen, Eric (April 17, 2017) *Physicists create 'negative mass.'* https://phys.org/news/2017-04-physicists-negative-mass.htm

Sowell, Thomas (1995) "Ethnicity and IQ". In Steven Fraser, ed., *The Bell Curve Wars.* New York: Basic Books, pp. 70-79.

SI Brochure. BIPM (December 22, 2013) Unit of time (second).

Spencer, Domina Eberle & Shama, Uma. *A New Interpretation of the Hafele-Keating Experiment.* http://www.shaping.ru/congress/english/spenser1/spencer1.asp

Stacks Physics Exchange: (2012) http://physics.stackexchange.com/questions/44934/does-matter-with-negative-mass-exist

Styer, Daniel F. (2011) *Relativity for the Questioning Mind.* Baltimore: Johns Hopkins University Press.

Susskind, Leonard (7 July 2008). *The Black Hole War: My Battle with Stephen Hawking to Make the World Safe for Quantum Mechanics*. Hachette Inc.

Susskind, Leonard (Jul 16, 2013) *Why is Time a One-Way Street?* Lecture at Santa Fe Institute. https://www.youtube.com/watch?v=jhnKBKZvb_U

Taylor, John (1980). *Science and the Supernatural: An Investigation of Paranormal Phenomena Including Psychic Healing, Clairvoyance, Telepathy, and Precognition by a Distinguished Physicist and Mathematician.* London: T. Smith.

Talbot, Michael (1992) *The Holographic Universe.* New York: Harper Collins.

Templeton, Graham (November 19, 2012) *Stanford's quantum entanglement device brings us one step closer to quantum cryptography.* Extreme Tech: www.extremetech.com/extreme/140739-stanfords-quantum-entanglement-device-brings-us-one-step-closer-to-quantum-cryptography

Tesla, Nikola (1932) *Space is Nothing.* New York Herald Tribune (September 11)

The Physics Classroom. *The Forbidden F-Word* (1996-2016). http://www.physicsclassroom.com/class/circles/Lesson-1/The-Forbidden-F-Word

Thornbill, Wallace and Talbott, David (2007) *The Electric Universe.* Portland, OR: Mikamar Publishing.

Trubody, Ben (June, 2017) "Richard Feyman's Philosophy of Science." *Philosophy Now*, Issue 120. https://philosophynow.org/issues/114/Richard_Feynmans_Philosophy_of_Science

Tylor, Edward B. (1877) *Primitive Culture.* New York: Henry Holt.

Tytell, David (April 13, 2004) "Building Planets In Plastic-Bags" http://www.skyandtelescope.com/astronomy-news/building-planets-in-plastic-bags/

University of Illinois Physics Website (2007) *Why are atomic masses not expressed as whole numbers?* https://van.physics.illinois.edu/qa/listing.php?id=1216

Veritasium YouTube site (2016) *The Absurdity of Detecting Gravitational Waves.* Interview with Prof Rana Adhikari. https://www.youtube.com/watch?v=iphcyNWFD10&feature=youtu.be

Vessot, R. F. C. and Levine, M. W. (1979, Dec.). Gravitational Redshift Space-Probe Experiment GP-A Project Final Report Contract NAS8-27969 Retrieved from http://ntrs.nasa.gov/archive/nasa/casi.ntrs.nasa.gov/19800011717.pdf.

Von Baeyer, Hans Christian (March 8, 2016) "Quantum Weirdness? It's All in Your Mind" in *Physics at the Limits*, Scientific American Publication.
Readers Respond to "Quantum Weirdness": https://www.scientificamerican.com/article/readers-respond-to-quantum-weirdness/

Wall, Mike (March 24, 2017) *Gravitational Waves Send Supermassive Black Hole Flying.* SPACE.com

Wasson, Valentina and Wasson, Gordon (1957) *Mushrooms, Russia and History.* New York: Pantheon, Vol. 2, 264-65.

Weber, Renee (1982) *The Holographic Paradigm* (Fritjof Capra interview) Shambhala/Random House, pp. 217–218.

Wolchover, Natalie (06.30.14) Have We Been Interpreting Quantum Mechanics Wrong This Whole Time? *Quanta Magazines.* https://www.wired.com/2014/06/the-new-quantum-reality/

Wigner, E.P. (1961), "Remarks on the mind-body question", in: I.J. Good, *The Scientist Speculates*, London, Heinemann.

Wikipedia (retrieved 2017) Faster-*than-Light.* https://en.wikipedia.org/wiki/Faster-than-light.

Wikiquotes: https://en.wikiquote.org/wiki/William_Thomson

Woit, Peter (2006). *Not Even Wrong: The Failure of String Theory and the Search for Unity in Physical Law.* Basic Books.

Yaglom, Isaak Moiseevich (1980) Mathematical Structures and Mathematical Modelling. Philadelphia: Gordon and Breach Science Publishers.

White, Leslie (1956) The Locus of Mathematical Reality. In: The World of Mathematics by Newman, James R. New York: Murray Printing Company.

Zato Tomáš (Jun 17, 2014) *What-Is-Hawking-Radiation-And-How-Does-It-Cause-A-Black-Hole-To-Evaporate.* Stack Exchange. https://physics.stackexchange.com/questions/26605/what-is-hawking-radiation-and-how-does-it-cause-a-black-hole-to-evaporate

Zukav, Gary. (1979) *The Dancing Wu Li Masters: An Overview of the New Physics.* HarperCollins. Kindle Edition.

www.ingramcontent.com/pod-product-compliance
Lightning Source LLC
Chambersburg PA
CBHW062336220526
45469CB00008B/2738